D0244012

Head Start to A-Level Chemistry

A-Level Chemistry is a **big step up** from GCSE... no doubt about that.
But don't worry — this CGP book has been lovingly made to help you **hit the ground running** at the start of your A-Level (or AS-Level) course.

It recaps everything you'll need to remember from GCSE, and introduces some of the crucial concepts you'll meet at A-Level. For every topic, there are **crystal-clear study notes** and plenty of **practice questions** to test your skills.

What CGP is all about

Our sole aim here at CGP is to produce the highest quality books — carefully written, immaculately presented and dangerously close to being funny.

Then we work our socks off to get them out to you — at the cheapest possible prices.

Contents

Section 7 — Organic Chemistry

Section 8 — Chemical Reactions

Section 9 — Rates of Reactions

Section 10 — Equilibria

Section 11 — Calculations

Section 12 — Enthalpy

Section 13 — Investigating and Interpreting

Published by CGP

Author:
David Mason

Editors:
Emily Howe and Sophie Scott

ISBN: 978 1 78294 280 1

With thanks to Katie Braid and Rachel Kordan for the proofreading.

Clipart from Corel®
Printed by Elanders Ltd, Newcastle upon Tyne.

Based on the classic CGP style created by Richard Parsons.

Text, design, layout and original illustrations © Coordination Group Publications Ltd. (CGP) 2015
All rights reserved.

Photocopying more than one chapter of this book is not permitted. Extra copies are available from CGP.
0800 1712 712 • www.cgpbooks.co.uk

Atomic Structure

What Are **Atoms** Like?

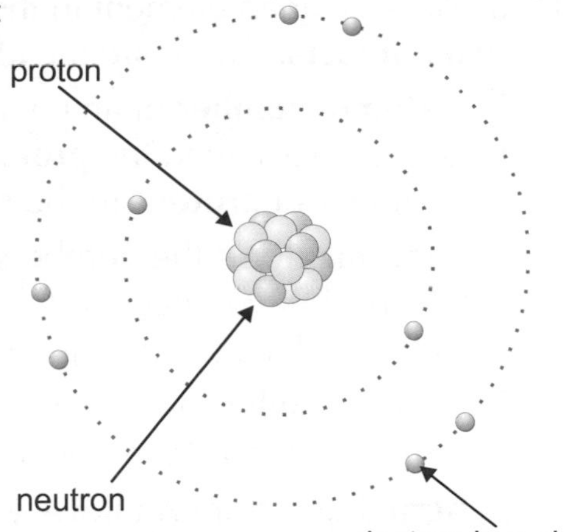

1) Atoms are made up of **three** types of **subatomic particle**: **protons**, **neutrons** and **electrons**.
2) In the **centre** of all atoms is a **nucleus** containing **neutrons** and **protons**.
3) Almost all of the **mass** of the atom is contained in the **nucleus** which has an overall **positive** charge. The positive charge arises because each of the **protons** in the nucleus have a **+1** charge.
4) The **neutrons** in the nucleus have a very similar **mass** to the protons but they are **uncharged**.
5) **Electrons** are much **smaller** and **lighter** than either the neutrons or protons. They have a **negative charge** (**–1**) and **orbit** the nucleus in **shells** (or energy levels).
6) There's an **attraction** between the **protons** in the nucleus and the **electrons** in the shells.
7) The nucleus is **tiny** compared with the total volume occupied by the whole atom.
8) The **volume** occupied by the **shells** of the electrons determines the **size** of the atom.

Here's a round up of the **properties** of the subatomic particles:

Particle	Relative Mass	Relative Charge
Proton	1	+1
Neutron	1	0
Electron	$\frac{1}{2000}$	–1

What is the **Charge** on an Atom?

The overall charge on an atom is **zero**.

This is because each **+1** charge from a **proton** in the nucleus is **cancelled out** by a **–1** charge from an **electron**.

If an atom **loses** or **gains** electrons it becomes **charged**. These charged particles are called **ions**.

EXAMPLE: How many electrons has an Al^{3+} ion lost or gained?

The Al^{3+} ion has a charge of **+3**, so there must be **3 more protons** than **electrons**.
Ions are formed when **electrons** are lost or gained, so Al^{3+} must have **lost 3 electrons**.

Neutrons are the perfect criminals — they never get charged...

1) Which subatomic particles are found in the nucleus?
2) What is the charge on an ion formed when an atom loses two electrons?
3) What is the charge on an ion formed when an atom gains two electrons?

Atomic Number, Mass Number and Isotopes

Atomic and Mass Numbers

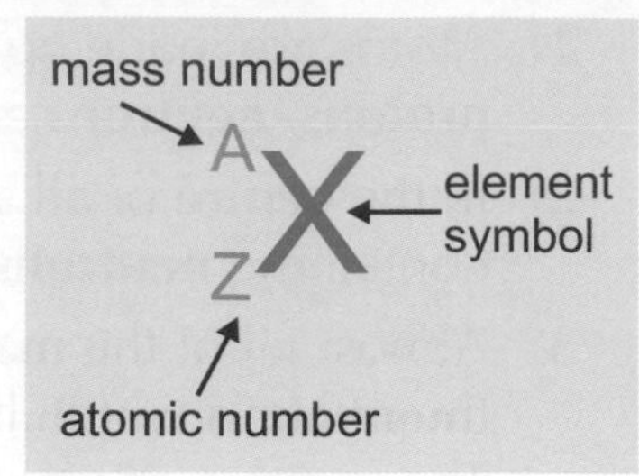

1) If you look at an element in the periodic table, you'll see it's given **two numbers**. These are the **atomic number** and the **mass number**.
2) The **atomic number** of an element is given the symbol **Z**. It's sometimes called the **proton number** as it represents the number of **protons** in the nucleus of the element.
3) For **neutral** atoms the number of **protons equals** the number of **electrons**, but you need to take care when considering ions as the number of electrons changes when an ion forms from an atom.
4) The **mass number** of an atom is given the symbol **A**. It represents the **total** number of **neutrons** and **protons** in the nucleus.
5) **Subtracting Z** from **A** allows you to calculate the number of **neutrons** in the nucleus.

EXAMPLE: Use the periodic table to complete the following information about sodium.

Element	Symbol	Z	A	No. Protons	No. Neutrons	No. Electrons
Sodium			23			

The periodic table tells you that the **symbol** for sodium is **Na** and **Z** is **11**.
The number of **protons** in sodium is the same as the **atomic number**, which is **11**.
You work out the number of **neutrons** by **subtracting** Z from A: 23 – 11 = **12**.
The number of **electrons** is the **same** as the number of protons, which is **11**.

Isotopes

1) Atoms of the same **element** always have the same number of **protons**, so they'll always have the same **atomic number**, but their **mass numbers** can **vary** slightly.
2) Atoms of the same **element** with different **mass numbers** are called **isotopes**.
3) Isotopes have the same number of **protons** but different numbers of **neutrons** in their nuclei.

EXAMPLE: Copper has an atomic number of 29. Its two main isotopes have mass numbers of 63 and 65. How many neutrons does each of the isotopes have?

The ^{63}Cu isotope has 63 – 29 = **34 neutrons**.
The ^{65}Cu isotope has 65 – 29 = **36 neutrons**.

Finding the number of neutrons — it's as easy as knowing your A – Z...

1) Use the periodic table to work out how many neutrons are in a neutral phosphorus atom.
2) In terms of the numbers of subatomic particles, state two similarities and one difference between two isotopes of the same element.
3) Three neutral isotopes of carbon have mass numbers 12, 13 and 14. State the numbers of protons, neutrons and electrons in each.

Relative Atomic Mass

Calculating the **Relative Atomic Mass**

1) The average mass of an element is called its **relative atomic mass**, or A_r.
2) When you look up the **relative atomic mass** of an element on a **detailed** copy of the periodic table, you'll see that it isn't always a **whole number**. This is because the value given is the **average** mass number of two or more **isotopes**.
3) The **value** of the relative atomic mass is further complicated by the fact that some isotopes are **more abundant** than others. It's a **weighted average** of all the element's different isotopes.
4) You can use the **relative abundances** and **relative isotopic masses** (the mass number of a single, specific isotope) of each isotope to work out the **relative atomic mass** of an element.
5) Relative abundances of isotopes are often given as **percentages**. To work out the **relative atomic mass** of an element, all you need to do is multiply **each isotopic mass** by its **relative abundance**, add all the values together and divide by **100**.

EXAMPLE: What is the relative atomic mass of chlorine given that 75% of atoms have an atomic mass of 35 and 25% of atoms have an atomic mass of 37?

Average mass = (abundance of $^{35}Cl \times 35$ + abundance of $^{37}Cl \times 37) \div 100$
$= [(75 \times 35) + (25 \times 37)] \div 100$
$= (2625 + 925) \div 100$
$= 3550 \div 100$
$= \mathbf{35.5}$ (You can check your answer against a periodic table to see if it's right.)

Calculating the **Relative Formula Mass**

If you **add up** the relative atomic masses of all the atoms in a chemical formula, you get the **relative formula mass**, or M_r, of that compound.

(If the compound is molecular, you might hear the term relative molecular mass used instead, but it means pretty much the same.)

EXAMPLE: Calculate the relative formula mass of $CaCl_2$.

Ca has an atomic mass of 40.1 and Cl has an atomic mass of 35.5.
$M_r = (1 \times 40.1) + (2 \times 35.5)$
$= \mathbf{111}$

Together, my brother and I weigh 143 kg — it's our relative mass...

1) Find the relative atomic mass of lithium if its composition is 8% ^{6}Li and 92% ^{7}Li.
2) Find the relative atomic mass of carbon if its composition is 99% ^{12}C and 1% ^{13}C.
3) Find the relative atomic mass of silver if its composition is 52% ^{107}Ag and 48% ^{109}Ag.
4) Find the relative formula mass of sodium fluoride, NaF.
5) Find the relative formula mass of chloromethane, CH_3Cl.

Electronic Structure

Electrons are Arranged in *Energy Shells*

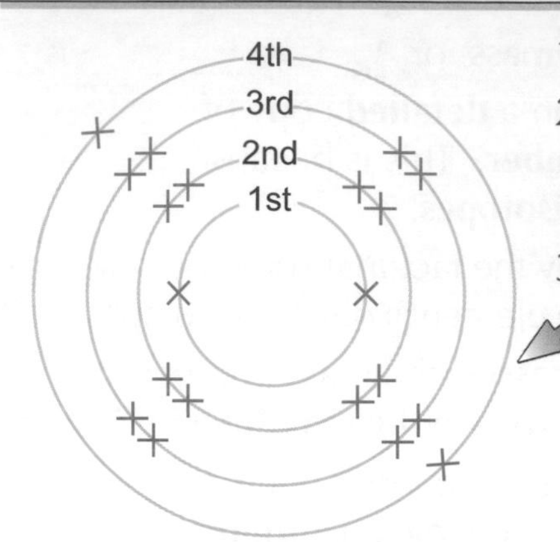

1) Electrons orbit the nucleus in **shells** (also called **energy levels**).
2) You can draw concentric **circles** to represent the different **shells**. Then add **crosses** to represent the **electrons** at each level.
3) For example, this diagram shows the energy levels, for an atom with 20 electrons, filling up with electrons. It has two electrons in the first shell, eight in the second shell, eight in the third shell and two in the fourth shell.
(Remember you should always **start** filling the **innermost levels** first.)

Here's another way to show electron arrangements using simple notation:

An atom with 6 electrons:	2, 4
An atom with 11 electrons:	2, 8, 1
An atom with 20 electrons:	2, 8, 8, 2

The first number tells you how many electrons are in the first shell, the second number tells you how many electrons are in the second shell, and so on.

Energy Levels are *Split* into *Subshells*

Subshell	Maximum electrons
s	2
p	6
d	10

1) Energy levels can be **split** into **sub-levels** called **subshells**. The first three subshells are called '**s**', '**p**' and '**d**'. They can each hold a different **number** of electrons.
2) The first energy level has **one subshell** — an 's' level. So the first energy level can contain up to **2 electrons**.
3) At GCSE you learnt that the second energy level can contain up to **8 electrons**. It's actually split into **2 sub-levels**. **Two** of the electrons are in an 's' level and the remaining **six** are in a '**p**' level. If you combine the 2 's' electrons with the 6 'p' electrons you get a total of 8.
4) Electrons generally start by filling the **energy level** with the **lowest energy**. So the **first** energy level will be completely **filled** before any electrons go into the **second** energy level. Within an energy level, electrons will fill the **subshells** in the order **s**, then **p**, then **d**.
5) As well as telling you how many electrons are in each **shell**, the **electron configuration** of an atom also tells you what **subshells** the electrons are in. For an atom with 10 electrons:

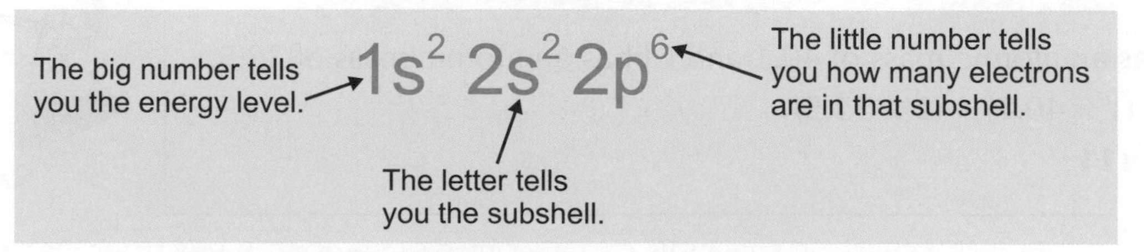

I'm trying to be calm, but my energy level is too high...

1) Draw diagrams to show the electron arrangements of the following elements: carbon, fluorine, magnesium, sulfur.
2) Use the simple notation shown above to write the electron arrangements of these elements: lithium, sodium, potassium, beryllium, magnesium, calcium.
3) Give the electron configurations of oxygen and chlorine.

The Periodic Table

The *Periodic Table*

The periodic table contains:

- All of the elements in order of atomic number.
- Vertical groups of elements which have similar properties.
- Horizontal rows of elements called periods.

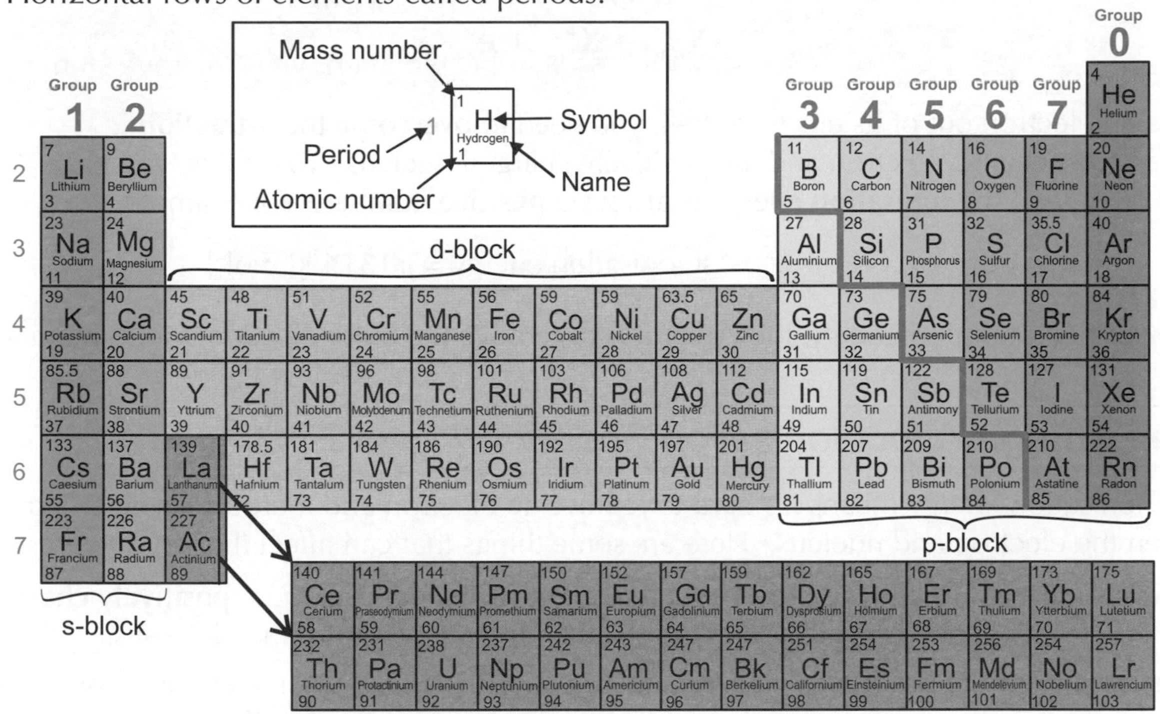

What Are **Groups** and **Periods**?

Chemical reactions involve atoms **reacting** to gain a **full outer shell** of electrons.
All of the elements in a group have the same **number** of **electrons** in their **outer shell**.
As a result, the elements in a group **react** in a **similar** way.

The properties of elements in the same **period change gradually** as you move from one side of the periodic table to the other.

The **Periodic Table** is Split into **Blocks**

1) As well as being split into groups and periods, the periodic table has four **blocks**. You only need to worry about two of them at the moment though — the '**s**' block and the '**p**' block.
2) Groups **1** and **2** are called the **s-block** elements. Their outer electrons are in energy levels called **s subshells**. S subshells can accommodate up to 2 electrons (see page 4).
3) Groups **3** to **0** are called the **p-block** elements. Their outer electrons are in energy levels called **p subshells**. P subshells can accommodate up to 6 electrons (see page 4).
4) There's always one exception. **Helium** (He) is an **s-block** element, even though it's in **Group 0**. Its electron configuration is $1s^2$.

The mystery of the periodic table? It's elementary, my dear Watson...

1) Sort the following elements into a table to show which ones are from the s-block, and which are from the p-block: caesium, potassium, phosphorus, calcium, aluminium, barium and sulfur.
2) Give one similarity between elements that are in the same group.

Ionisation Energy

When Atoms **Lose Electrons** they are **Ionised**

When electrons have been removed from an atom or molecule, it's been **ionised.**
The energy you need to remove the **first outer electron** is called the **first ionisation energy**.

The first ionisation energy is the energy needed to **remove 1 electron** from **each atom** in **1 mole** of **gaseous** atoms to form 1 mole of gaseous 1+ ions (see page 37 for more on moles).

$$X_{(g)} \rightarrow X^+_{(g)} + e^-$$

To take an electron out of its electron shell, you need to **overcome** the **attraction** between the negative electron and the positively charged nucleus. To do this, you have to **add** energy, so the **ionisation energy** is always a **positive number**. For example:

$O_{(g)} \rightarrow O^+_{(g)} + e^-$ 1st ionisation energy = +1314 kJ mol^{-1}

The **lower** the ionisation energy, the **easier** it is to remove the outer electron and form an ion.

***Three Things** Affect Ionisation Energy*

A high ionisation energy means it's **hard** to remove an electron and there's a **stronger** attraction between the electron and nucleus. Here are some things that can affect the ionisation energy:

1) **Nuclear charge**: The **more protons** there are in the nucleus, the more **positively charged** the nucleus is and the **stronger** the attraction for the electrons.
2) **Distance from the nucleus**: Attraction decreases with **distance**. An electron **close** to the nucleus will be more strongly attracted than one **further away**.
3) **Shielding**: Electrons in shells **closer** to the nucleus can **stop** the outer electrons from feeling the **full force** of the nuclear charge. The inner electrons are said to **shield** the outer electrons from the nucleus. More inner electrons mean more shielding, so a **weaker attraction** for the **outer electrons** and a **lower ionisation energy**.

*The Periodic Table Shows **Trends** in Ionisation Energies*

1) Ionisation energy **decreases** down a **group**. This is because as you go down a group, each element has **one more** electron shell than the one above — so the distance between the **nucleus** and the **outer shell increases**. There will also be more **shielding** from the larger number of **inner electrons**. So overall, going down a group the **attraction** between the nucleus and the outer electrons **decreases**.
2) Ionisation energy generally **increases** across a **period**. There are **more protons** in the nucleus so there's a **higher nuclear charge**. Electrons are also going into the **same shell**, so the **distance** from the nucleus and the amount of **shielding** by inner electrons doesn't change much. So overall, the attraction between the nucleus and the electrons **increases**.

Let it go, let it go, lose electrons from my outer shell...

1) Write an equation to show the first ionisation of sodium.
2) What three things can affect ionisation energy?
3) For the following pairs of elements, decide which will have the higher first ionisation energy: Magnesium and Calcium, Lithium and Fluorine, Oxygen and Sulfur.

Formation of Ions

Elements in the **s-block** and the **p-block** form **Simple Ions**

Most elements in the **s-block** and the **p-block** form ions with **full outer electron shells**. This means you can **predict** what ion an element will form by looking at the **periodic table** — just follow through the reasoning below:

- **Group 1** atoms have **one electron** in their outer shell. The **easiest way** for them to achieve a full outer shell is to **lose** that one negative electron. The positive charge in the nucleus stays the same leaving one excess positive charge overall, so **Group 1 ions** must have a **1+ charge**.
- **Group 2** atoms have **two electrons** in their outer shell. They **lose** these two negative electrons to get a **stable** (full) outer shell, producing ions with a **2+** charge.
- **Group 6** elements have six electrons in their outer shell. Rather than releasing all six of these electrons (which would take a lot of energy) they **pick up** two electrons from their surroundings to complete their outer shell. The positive charge in the nucleus stays the same, so Group 6 ions have **two extra** negative charges — they carry a **2– charge**.
- **Group 7** atoms need to pick up **one** extra electron to get a stable outer shell, so they form ions with a charge of **1–**.

Generally the charge on a **metal ion** is equal to its **group number**.
The charge on a **non-metal ion** is equal to its **group number minus eight**.

Not all **Ions** are as **Simple**

Some **groups of atoms** can also exist as stable ions. These are usually **anions** (negative ions) like sulfate and carbonate (one of the few exceptions being ammonium with a 1+ charge). It is harder to work out the charges on these than in the case of the simple ions above.

It is useful to **learn** the charges on the most common of these **molecular ions**:

+1	–2	–1
NH_4^+ (ammonium)	SO_4^{2-} (sulfate)	OH^- (hydroxide)
	CO_3^{2-} (carbonate)	NO_3^- (nitrate)
	SO_3^{2-} (sulfite)	HCO_3^- (hydrogencarbonate)
		CN^- (cyanide)

Transition metals (the block of elements between Groups 2 and 3) also form ions. They are **positive** (like all metal ions) but they **do not** form ions with a full outer shell of electrons. This means you can't predict the charges in the same way as you can with the s-block metals.

Most transition metals can form **more than one** ion. The different charges are called '**oxidation numbers**' of the element (see page 8). The common ones that you should be aware of are:

Fe^{2+}, Fe^{3+}, Cu^{2+}, Co^{2+}, Ni^{2+}, Zn^{2+} and Cr^{3+}

I never ask for an anion's opinion — they're always so negative...

1) What is the charge on a sodium ion?
2) Which Group typically forms 1– ions?
3) What is the formula of a sulfite ion? Remember to include the overall charge on the ion.

Oxidation Numbers

Oxidation Numbers Tell you the *Charge* on an *Atom*

When atoms **react** or **bond** to other atoms, they can **lose** or **gain** electrons.

The **oxidation number** tells you how many electrons an atom has donated or accepted when it's reacted. You may also see **oxidation numbers** called **oxidation states**, but they're the same thing.

Roman Numerals Tell you the *Oxidation Number*

Roman numerals can be used to show what oxidation number a certain element has. You'll probably remember your Roman numerals from Maths, where (I) = +1, (II) = +2, (III) = +3 and so on. The Roman numerals are written **after** the name of the **element** they correspond to.

In iron(II) chloride, iron has an oxidation number of +2. Formula = $FeCl_2$
In iron(III) chloride, iron has an oxidation number of +3. Formula = $FeCl_3$

There are Some *Rules* About *Oxidation Numbers*

1) Elements that aren't bonded to anything else all have an oxidation number of **0**.

Ne Ag

Uncombined elements.
Oxidation number = 0

2) Elements that are bonded to identical atoms also have an oxidation number of **0**.

F F N N

Elements bonded to identical elements.
Oxidation number = 0

3) The oxidation number of an ion made up of just one atom is the same as its **charge** (the little number to the **right** of the symbol).

4) For **molecular ions** (ions that are made up of more than one atom) the **overall charge** of the whole ion is equal to the **sum** of the **oxidation numbers** of the individual atoms or ions.

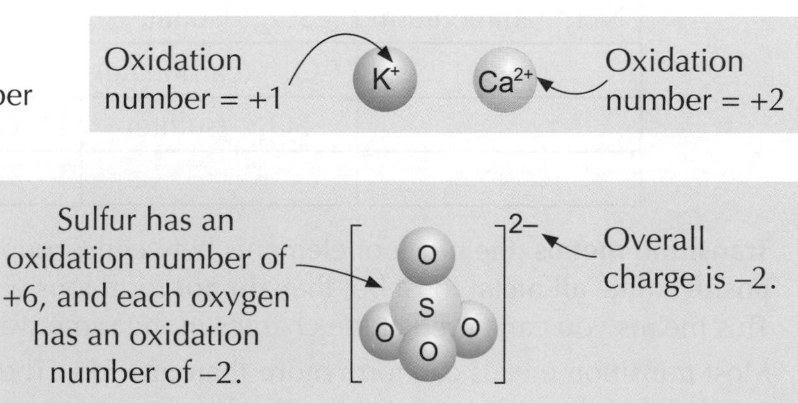

Caesar the day — and get to know your Roman numerals...

1) What does the oxidation number tell you about an atom?
2) Give the oxidation number of each of the following atoms/ions: Al^{3+}, H^+, Ne, O^{2-}.
3) What is the oxidation number of an atom of chlorine in Cl_2?

Intermolecular Bonding

*Intermolecular Bonds Form Between **All** Molecules*

1) Some compounds are made up of **simple molecules** — these are just groups of a **few atoms** joined together by **covalent bonds** (see page 13). For example, water (H_2O) or oxygen (O_2).
2) The bonds **between** the **atoms** in each molecule are very **strong**. By contrast, there are very **weak** forces of attraction that form **between** the **molecules**. These are **intermolecular bonds** (also called intermolecular forces).

Weak intermolecular forces

Strong covalent bond

*The **Strength** of Intermolecular Bonds Affects **Boiling Points***

When simple molecular substances **melt** or **boil**, it's the **intermolecular bonds** that get broken — not the much stronger covalent bonds. The **stronger** the intermolecular bonds, the more **energy** is needed to break them, so the **higher** the boiling or melting point will be. Two things that can affect the strength of intermolecular forces are:

1) The number of **electrons** in a molecule: the **more** electrons there are, the **stronger** the intermolecular bonds between molecules.
2) The **surface area** of the molecule: the **larger** the surface area over which intermolecular bonds can act, the **stronger** the intermolecular bonds between molecules.

EXAMPLE: Use the idea of intermolecular bonds to explain the trend in boiling points of the following alkanes.

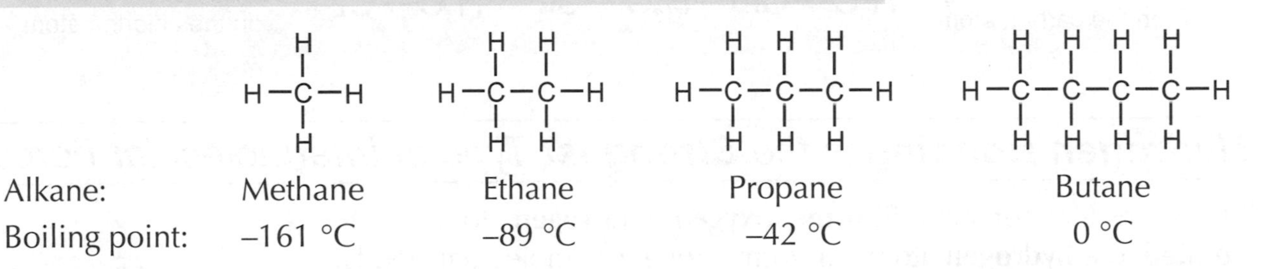

Alkane:	Methane	Ethane	Propane	Butane
Boiling point:	–161 °C	–89 °C	–42 °C	0 °C

There is a clear trend showing that as the molecules get **larger** their boiling point **increases**.

This is due to the fact that the larger molecules have a greater **surface area**, so there is stronger intermolecular bonding. The larger molecules also have more **electrons** — this further increases the strength of the intermolecular bonds that form between molecules.

This page was alright — we formed a sort of bond...

1) Draw a diagram to show the different types of bonding in a sample of gaseous chlorine molecules ($Cl_{2(g)}$). What type of bond is the strongest?
2) Use the data in the example above to predict the boiling points of the next four members of the alkane series. They are called pentane, hexane, heptane and octane. (Bear in mind that, in Chemistry, the first member of a series does not always provide an ideal example.)

Polarity

Some Atoms Attract **Bonding Electrons** *More* **Strongly**

The ability of an atom to **attract** electrons in a **covalent bond** is called its **electronegativity**.

1) Ignoring Group 0, electronegativity **decreases** down a **group** in the periodic table, and **increases** across a **period**. The **most** electronegative element is **fluorine**.
2) In a bond between two **different** elements with different **electronegativities**, the **bonding electrons** will be attracted **more strongly** towards the atom with the **higher** electronegativity. This makes the bond **polar**.

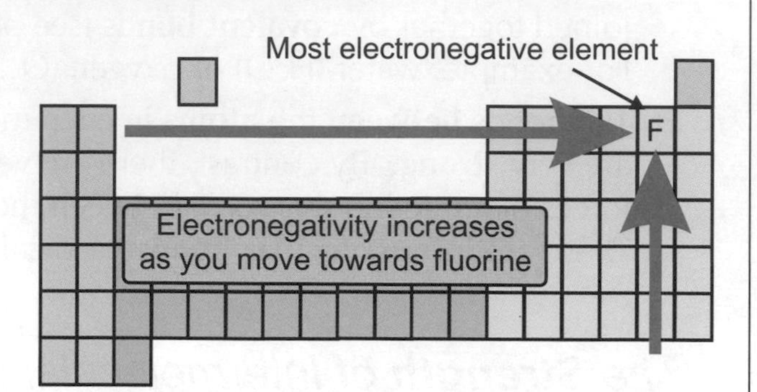

Polar Bonds *Can Affect the* **Strength** *of* **Intermolecular Forces**

If you substitute a chlorine atom for one of the hydrogen atoms in a methane molecule, it has a marked effect on the boiling point.

Boiling point of methane (CH_4)	= –161 °C
Boiling point of chloromethane (CH_3Cl)	= –24 °C

The reason for the dramatic increase in boiling point is that the chlorine atom **polarises** the molecule, making one end **slightly positive** and the other **slightly negative**. The **oppositely charged** ends of **different** molecules **attract** each other, so more energy is required to separate them. This results in an increase in boiling point.

'δ+' means there's a small positive charge on the carbon atom

$\overset{\delta+}{H_3C}—\overset{\delta-}{Cl}\cdots\overset{\delta+}{H_3C}—\overset{\delta-}{Cl}\cdots\overset{\delta+}{H_3C}—\overset{\delta-}{Cl}$

'δ–' means there's a small negative charge on the chlorine atom

Hydrogen Bonding *is the* **Strongest** *Type of Intermolecular Force*

Molecules that contain a **fluorine**, **oxygen** or **nitrogen** atom **bonded** to a **hydrogen** atom can form strong intermolecular bonds.

This is because the hydrogen atoms are strongly **polarised** by the very electronegative fluorine, oxygen or nitrogen atoms. These slightly positive hydrogen atoms are attracted to the lone pair of electrons on a F, O or N atom in a **nearby molecule** to form an attraction known as a **hydrogen bond**.

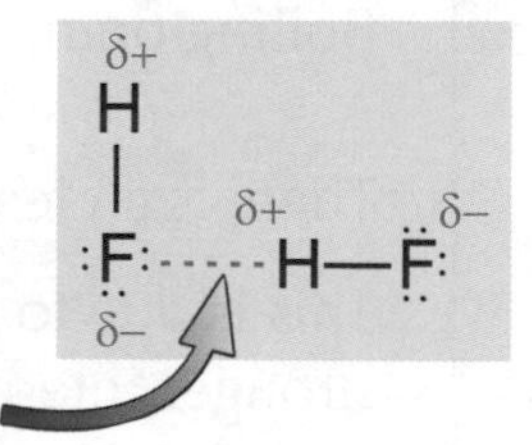

Hydrogen bonds are the strongest type of intermolecular attraction, though they are not as strong as either an ionic or a covalent bond (see next section).

Polar Bond — the Arctic's answer to 007...

1) For the following pairs of molecules, predict with reasoning which has the higher boiling point:
   a) H_2 and HF, b) H_2O and H_2S, c) CH_3F and CH_3I.
2) Water is a polar molecule. Draw a diagram showing three water molecules attracted together. You should use dotted lines to indicate forces between atoms in different molecules. The shape of a water molecule is shown on the right.

Ionic Bonding

Ionic Bonds Involve the Transfer of Electrons

1) Ions form when **electrons** are transferred from **one atom** to **another**. Atoms that **lose electrons** form **positive ions** and atoms that **gain electrons** become **negative ions**.
2) These oppositely charged ions are **attracted** to each other by **electrostatic attraction**. When this happens, an **ionic bond** is formed.
3) The simplest ions form when atoms lose or gain 1, 2 or 3 electrons to get a **full outer shell**.
4) You can show the transfer of electrons to form an ionic compound using a **dot-and-cross** diagram. For example, sodium and chlorine will react to form sodium chloride (NaCl):

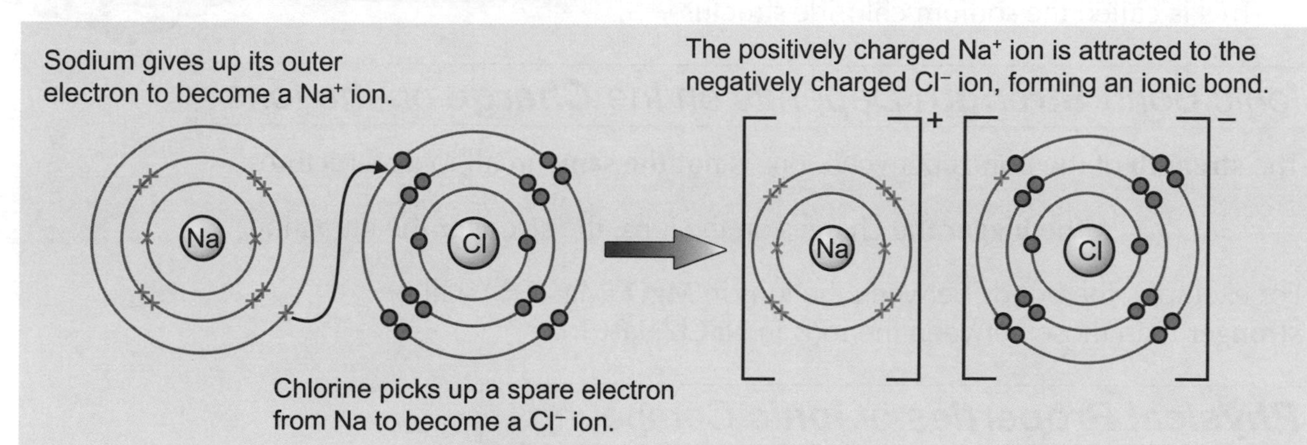

5) In the example above, the **dots** represent the electrons that come from the chlorine atom, and the **crosses** represent the electrons that come from the sodium atom.

You Can Find The Ratio of Positive to Negative Ions

1) The **ratio** of positive ions to negative ions in an ionic compound depends on the **charges** of the ions.
2) The **overall charge** of an ionic compound is **zero**, so the **sum** of all the **positive charges** in the compound must be **equal** to the **sum** of the **negative charges**.
3) If you know the **individual charges** of each of the ions in a compound, you can work out their **ratio**. You can use this to find the **ionic formula** of the compound.

EXAMPLE: In the compound calcium chloride, what is the ratio of Ca^{2+} to Cl^- ions?

For the compound to be neutral it must contain **two Cl^-** ions (2 × −1) to **balance** the charge of **each Ca^{2+}** ion (1 × +2).
So the ratio of Ca^{2+} ions to Cl^- ions in the compound must be **1 : 2**.
The ionic formula will be **$CaCl_2$**.

I can't afford Mg^{2+} — the charge is just too high...

1) Draw a diagram showing how a magnesium atom reacts with an oxygen atom to form magnesium oxide, MgO. Your diagram should show the electron transfer process.
2) In potassium oxide, what is the ratio of K^+ ions to O^{2-} ions? What is the ionic formula?

Ionic Compounds

Ionic Bonds Produce *Giant Ionic* Structures

1) Ionic bonds do not work in any particular direction.
   The electrostatic attraction is just as strong in **all directions** around the ion.
2) This means that when ionic compounds form, they produce **giant lattices**.
3) The lattice is a closely packed **regular** array of ions, with each negative ion **surrounded** by positive ions and vice versa.
   The **forces** between the **oppositely charged** ions are very **strong**.
4) **Sodium chloride** forms a lattice like this one.
   This is called the sodium chloride structure.

Ionic Bond **Strength** *Depends on the* **Charge** *on the Ions*

The **strength** of the bonds between ions is **not the same** in all ionic structures:

The **bigger** the charges on the ions, the **stronger** the attraction.

For example, the bonds between the ions in **MgO** ($Mg^{2+}O^{2-}$) will be **stronger** than those between the ions in **NaCl** ($Na^{+}Cl^{-}$).

Physical Properties of *Ionic* Compounds

Melting points

In order to **melt** a solid, the forces holding the particles together have to be **overcome**. In an ionic solid, these bonds are very **strong**, so a **large** amount of energy is required to break them. So, ionic compounds have very **high** melting points.

Electrical conductivity

In their solid form, ionic compounds are electrical **insulators** (they don't conduct electricity). They have **no free ions** or electrons to carry electric current.
When **molten** or **dissolved**, the ions **separate** and are **free** to move and conduct electricity.
So **all** ionic compounds **conduct** electricity when **molten** or **dissolved**.

Solubility

In many cases ionic compounds are **soluble** in water.
This happens because water is a **polar** molecule (see page 10) — the positive end of the molecule points towards the negative ions and the negative end towards the positive ions.
Although **lots of energy** is required to break the strong bonds within the lattice, it is provided by the formation of **many weak bonds** between the water molecules and the ions in solution.

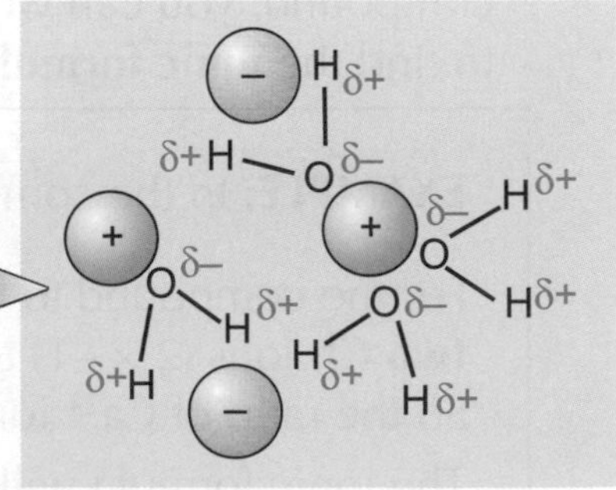

Rabbits love studying ionic compounds — all those giant lettuces...

1) Put these ionic compounds in order of melting point, highest to lowest: Lithium oxide (Li_2O), Beryllium oxide (BeO), Lithium fluoride (LiF). Explain why you have put them in that order.
2) Explain why the ionic compound, potassium chloride (KCl), can conduct electricity when molten or dissolved, but not when it is solid.

Covalent Bonding

Covalent Bonding *Involves* ***Shared Pairs*** *of Electrons*

Ionic bonding only really works between elements that have to gain or lose one or two electrons to get a full outer shell. Elements with **half-full** shells have to do something different. These elements **share** their electrons with another atom so they've both got a full outer shell. Both positive nuclei are **attracted** to the shared pair of electrons.
This results in the formation of **covalent bonds**.

A covalent bond is a **shared pair** of electrons.

For example, two hydrogen atoms share a pair of electrons to form a covalent bond:

H H → H H ← The shared pair of electrons means that each hydrogen atom in the molecule has 2 electrons in its outer shell.

When a small number of atoms share electrons in this way, a small covalent molecule forms. Such molecules can be represented in several different ways:

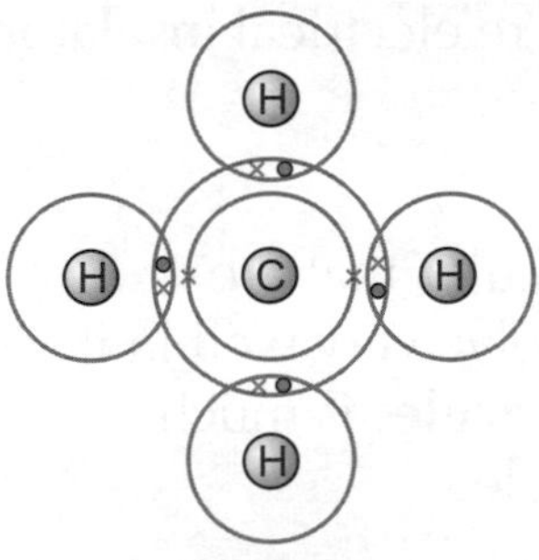

Dots represent electrons from the Hs and crosses represent electrons from C.

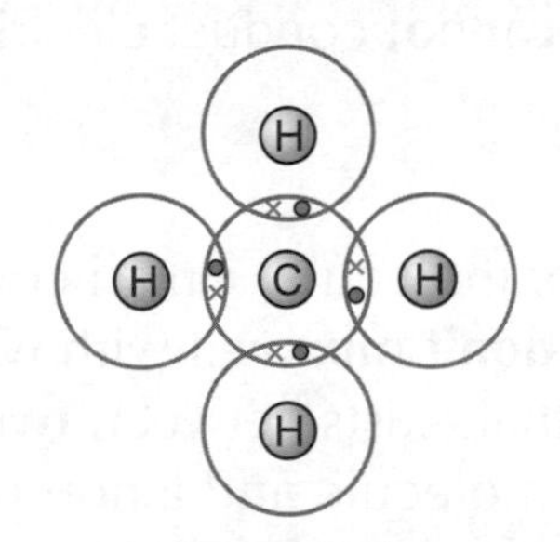

A more simple dot-and-cross diagram, showing only the outer shells of electrons.

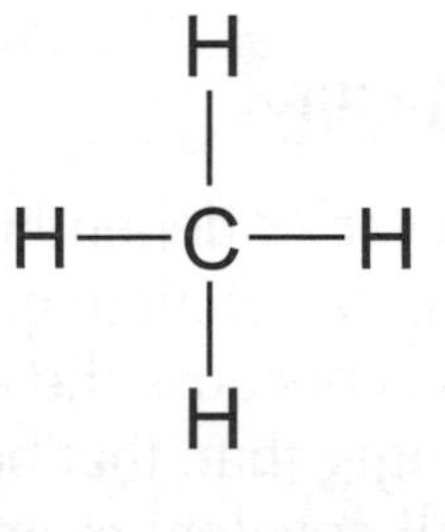

Each dash represents a single covalent bond (this is the most common notation).

If two atoms share **more than one** pair of electrons between them, then a **multiple covalent bond** can form. For example, in **carbon dioxide** (CO_2), there are two C=O double bonds:

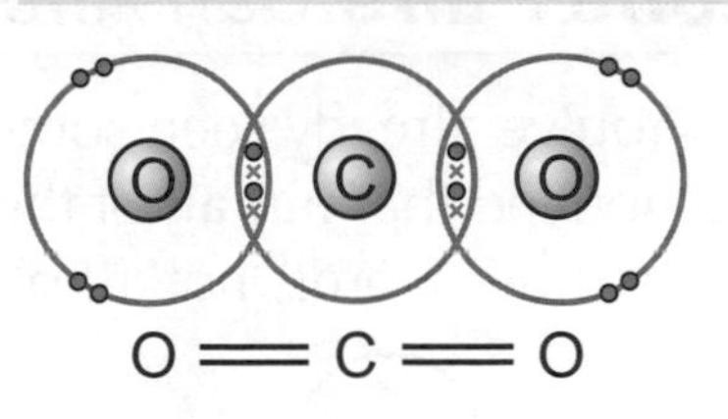

Dative *Covalent Bonding*

In **dative covalent bonds**, **both** of the shared electrons in the covalent bond come from the **same atom**. For example, in the ammonium ion (NH_4^+) there is a dative covalent bond formed from the nitrogen to a hydrogen ion (H^+):

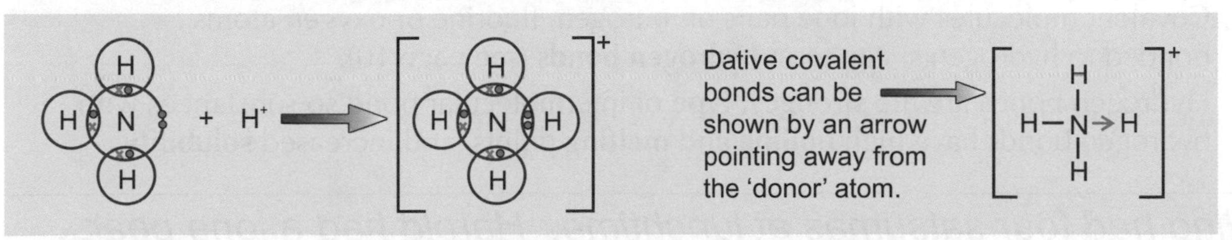

Friendly atom with GSOH WLTM special someone to share a bond with...

1) Draw simple 'dot-and-cross' diagrams to show the bonding in the following molecules:
a) chlorine (Cl_2) b) water (H_2O) c) ethane (C_2H_6) d) oxygen (O_2)

Small Covalent Molecules

Properties of Small Covalent Compounds

Small covalent compounds are made up of **lots** of small covalent molecules.
There are **strong covalent bonds** between the **atoms** in each molecule, but very **weak**, **intermolecular bonds** between the individual molecules (see page 9). It is these intermolecular bonds that determine the physical properties of small covalent compounds.

Melting points

In order to **melt** (or boil) a small covalent compound, you just have to break the **weak** intermolecular bonds **between** the molecules (not the strong covalent bonds). This doesn't need much energy, so small molecules normally have very **low** melting and boiling points — they're often liquids (e.g. water, H_2O) or gases (e.g. oxygen, O_2) at room temperature.

Electrical Conductivity

Small covalent molecules don't contain any of the **free charged** particles that are needed to carry an electric current. As a result they **cannot** conduct electricity — they're electrical **insulators**.

Solubility

This **varies** depending on the **type** of molecule. Small covalent molecules that are **not polar** at all (e.g. hydrocarbons) **don't mix** well with water, or dissolve very well in it. This is because the attractive force that exists between **two water molecules** is much **stronger** than that between a water molecule and a non-polar molecule.
Small covalent molecules that are **polar** or can form **hydrogen bonds** (see page 10) **can** dissolve in water.

Lone Pairs Can Affect the **Physical Properties**

1) You've already seen some examples of small covalent molecules on page 13. You may have noticed that **not all** of the electron pairs around the central atom are bonding electrons. In other words, not all of the electrons are **shared** between the atoms in the molecule.

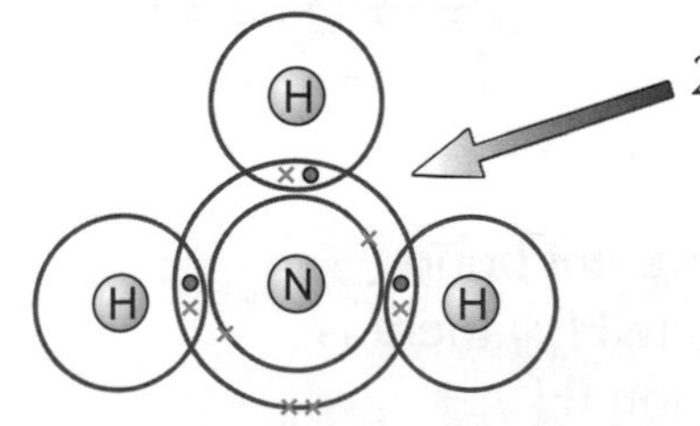

2) In ammonia (NH_3) there are **4 electron pairs** around the central nitrogen atom. **Three** of these electron pairs are called **bonding pairs** as they are **shared** between the **nitrogen** and **hydrogen atoms**. The **fourth** electron pair is **not shared** between the atoms in the molecule. This is called a **lone pair**.
3) Covalent molecules with lone pairs on nitrogen, fluorine or oxygen atoms, bonded to hydrogen(s) can form **hydrogen bonds** (see page 10).
4) Hydrogen bonds are the **strongest type** of intermolecular bond so substances with hydrogen bonds have high **boiling** and **melting** points, and increased **solubility**.

Aisling had four satsumas at lunchtime. Harold had a lone pear...

1) Draw a dot-and-cross diagram to show the bonding in hydrogen fluoride (HF). Label the bonding electrons and lone pairs of electrons.
2) Explain why nitrogen is a gas at room temperature, despite the nitrogen atoms in each molecule being strongly bonded to each other.

Giant Covalent Structures

Giant Covalent Structures

Carbon is ideally placed to share electrons and form covalent bonds, because it has a **half-full** outer shell. Carbon atoms can share their electrons with four other carbons to gain a full outer shell. This can result in the formation of a single massive carbon molecule — a **giant structure**. Carbon can form various different **giant covalent structures** such as **diamond** and **graphite**.

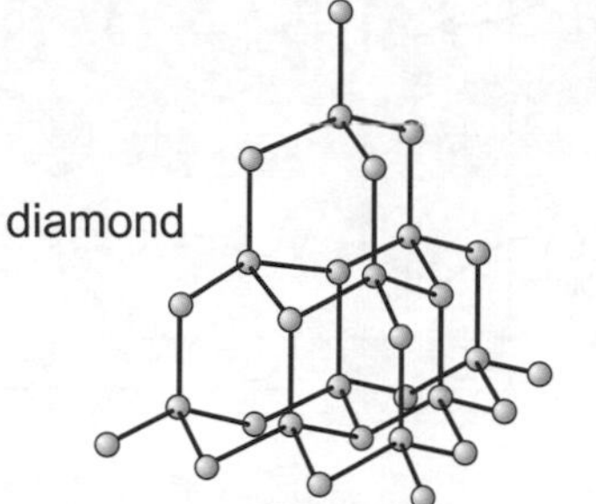

Each carbon atom forms **four** covalent bonds in a very **rigid** structure.
This structure makes diamond very **hard**.

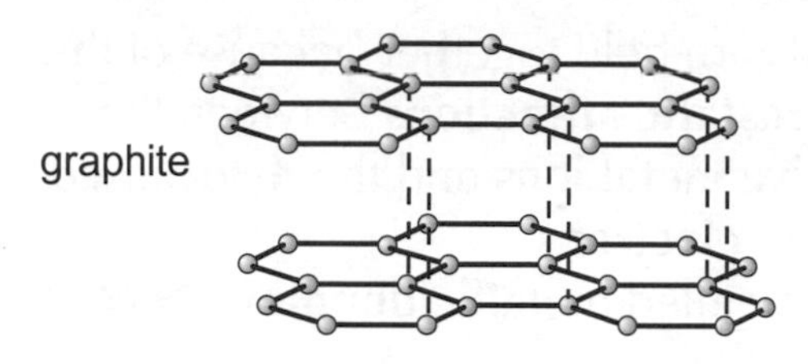

Each carbon atom forms **three** covalent bonds in the same **plane**. This results in a series of **layers** which can **slide** over each other.
The fourth electron from each carbon atom is **free**.

Properties *of Giant Covalent Structures*

Giant covalent structures have some different **physical properties** from small molecules.

Melting points

Unlike small molecules, melting points are **extremely high**, as all of the atoms are held together by **strong covalent bonds**. These millions of covalent bonds need to be **broken** to allow the atoms within the structure to move freely, which requires a lot of energy.

This contrasts with small molecules where no covalent bonds (only intermolecular bonds) need to be broken in order for the substance to melt.

Electrical conductivity

Giant covalent structures are **electrical insulators**. This is because they don't contain **charged particles**, and the atoms aren't free to move.
Even a **molten** covalent compound will not conduct electricity.
Graphite is the only exception to this, as the loosely held **electrons** between the layers of atoms can move through the solid structure. Graphite conducts in both its solid and liquid forms.

Solubility

Giant covalent structures are **not soluble** in water. To get a giant covalent structure to dissolve, all the covalent bonds joining the atoms together would need to be **broken**. There is no way to get the energy required to do this, since the individual **neutral atoms** in the structure will **not** form intermolecular bonds with the water molecules.

Diamonds — don't mess with 'em — they're well 'ard...

1) Devise a series of tests that would allow you to distinguish between two unknown crystalline solids, one of which is an ionic compound and the other a giant covalent structure.
2) Why won't diamond dissolve in water when sodium chloride will?

Metallic Bonding

Metals have Giant Structures Too

1) In a metal, the **outer electrons** from each atom are **delocalised** (they're not stuck on one atom) — this leaves **positive metal ions**.
2) The positive metal ions are arranged regularly in a **giant structure**, surrounded by a 'sea' of delocalised electrons.
3) Metals are held together because of the **electrostatic attractions** between the **positive metal ions** and the **delocalised 'sea'** of **electrons**.
   This is called **metallic bonding**.

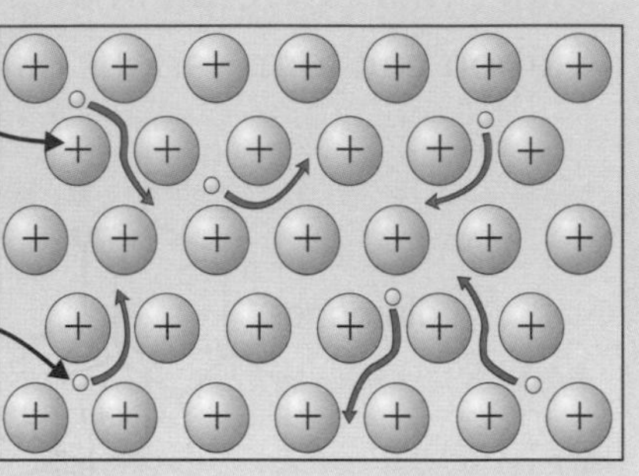

Properties of Metals

Metallic bonding explains the **physical properties** of metals:

Melting points

Metals generally have **high** melting points. This is because a lot of energy is required to overcome the **strong metallic bonding** between the particles.

The **more** electrons that are **delocalised** from **each atom**, the **stronger** the bonding will be and the **higher** the melting point.

EXAMPLE: Predict, with reasoning, whether magnesium or sodium will have a higher melting point.

Magnesium is made up of **Mg^{2+}** ions with **two** delocalised electrons per atom. Sodium is made up of **Na^{+}** ions and only **one** delocalised electron per atom.

So **magnesium** will have a **higher melting point** than sodium, because the metallic bonds will be **stronger** and require **more energy** to break.

Electrical conductivity

The **delocalised electrons** in metals are **free to move** around and can carry a **current**. This makes metals **good electrical conductors**.

Solubility

The **strong metallic bonds** mean that metals are generally **insoluble**.

Metallica bonds — friendships based on a love of '80s rock music...

1) Predict, with reasoning whether potassium or calcium will have a higher melting point.
2) Draw a diagram to show the bonding in a sample of sodium.
3) Sodium has a metallic structure, whilst sodium chloride (NaCl) is an ionic compound. Give one similarity and one difference between the physical properties of these substances.

Trends in Properties Across the Periodic Table

***Structure** and **Bonding Change** Across the Periodic Table*

You should have seen from this section how much the **properties** of a compound depend on its **bonding** and **structure**. You also know that the type of bonding that occurs depends on the **number of electrons** in the outer shells of the elements making up the compound, and so their **positions** in the periodic table.

A good way to compare the way that different elements bond is by looking at the properties of a series of similar compounds across a **period**.
Look at the information in the table below about all the Period 3 oxides.
You can see that there are clear **patterns** in the data.

Trends** Across **Period Three

(Period 3 is studied because it is a simple case.
There are no d-block elements to confuse matters.)

The table below shows some of the physical properties of Period 3 oxides.
The final row has been deduced from these physical properties.

	Na_2O	MgO	Al_2O_3	SiO_2	P_4O_{10}	SO_2
State (at room temperature and standard pressure)	solid	solid	solid	solid	solid	gas
Melting point (°C) (at standard pressure)	1275	2800	2072	1650	570	–73
Electrical conductivity (when molten)	good	good	good	none	none	none
Bonding	ionic lattice	ionic lattice	ionic lattice	giant covalent structure	small covalent molecule	small covalent molecule

You can see from this data that there is a change in the properties of the Period 3 oxides as you move from left to right across the periodic table.

The trend is from **ionic** bonding to **small covalent** molecules via a **giant covalent structure**.

These trends across a period are **more subtle** than the trends going down a group that you saw at GCSE. However they are extremely useful as they allow you to make **predictions** about the reactions and properties of unknown compounds.
There are of course **exceptions** to the rules/trends, but on the whole they allow links between physical properties and atomic structure to be made.

Trending now — #Arewedoneyet? #Dontworryitstheendofthesection...

1) Explain how the data in the first three rows of the table above supports the idea that the bonding type changes from ionic to covalent as you move across Period 3.
2) Use the information on Period 3 oxides to predict the trend in the melting points of the elements as you go across Period 3.
3) Predict the type of bonding you would expect in the chlorides of:
   a) sodium   b) phosphorus

Writing and Balancing Equations

Reaction Equations Show How Chemicals React Together

1) A reaction equation shows what happens during a chemical reaction. The **reactants** are shown on the **left hand side**, and the **products** on the **right hand side**.
2) **Word equations** just give the **names** of the components in the reaction. For example: propane + oxygen → carbon dioxide + water
3) **Symbol equations** give the chemical formulae of all the different components. They show all the **atoms** that take part in the reaction, and how they rearrange. For example: $C_3H_8 + 5O_2 \rightarrow 3CO_2 + 4H_2O$
4) Symbol equations have to **balance** — there has to be the **same number** of each **type** of atom on each side of the equation. The big numbers in front of each substance tell you how much of that particular thing there has to be for all the atoms to balance.

Writing **Balanced** Equations

To write a balanced symbol equation for a reaction there are 4 simple steps:

1) Write out the **word equation** first.
2) Write the correct **formula** for each substance below its name.
3) Go through each element in turn, making sure the **number of atoms** on each **side** of the equation **balances**. If your equation isn't balanced, you can only add more atoms by adding **whole reactants** or **products**.
4) If you changed any numbers, do step 3 again, and repeat until **all** the elements **balance**.

Doing the third step:

If the atoms in the equation don't balance you **can't** change the **molecular formulae** — only the numbers in **front** of them.

For example: $CaO + HCl \rightarrow CaCl_2 + H_2O$

There are **two Cl** atoms on the **right-hand side** of the equation, so we need to have **two HCl** on the **left-hand** side: $CaO + \mathbf{2}HCl \rightarrow CaCl_2 + H_2O$

This also doubles the number of **hydrogen atoms** on the left-hand side, so that the hydrogens **balance** as well.

EXAMPLE: Write a balanced equation for the reaction of magnesium with hydrochloric acid.

Step 1 — Write the word equation:
magnesium + hydrochloric acid → magnesium chloride + hydrogen

Step 2 — Write the symbol equation: $Mg + HCl \rightarrow MgCl_2 + H_2$

Step 3 — Go through the equation and balance the elements one by one:
$Mg + \mathbf{2}HCl \rightarrow MgCl_2 + H_2$
(the Mgs balance, but there are different amounts of H and Cl on each side. Put a 2 in front of HCl to balance the Hs and Cls. Check everything still balances.)

Writing and Balancing Equations

In **Ionic Equations** Make Sure the **Charges** Balance

1) In some reactions, particularly those in solution, not all the particles take part in the reaction.
2) **Ionic equations** are chemical equations that just show the **reacting particles**.
3) As well as having the same number of **atoms** of each element on each side of the equation, in ionic equations you need to make sure the **charge** is the same on both sides.

EXAMPLE: Balance the following ionic equation: $Na + H^+ \rightarrow Na^+ + H_2$

First, balance the **number of atoms** of each element using the method on the last page:

$$Na + \mathbf{2}H^+ \rightarrow Na^+ + H_2$$

Then check the **charge** is the same on both sides of the equation:

- On the left hand side, each H^+ ion contributes +1, so the charge is 2 × +1 = **+2**.
- On the right hand side, the sodium ion contributes +1, so the charge is 1 × +1 = **+1**.

To get the charges to balance, you need another positive charge on the right-hand side. One way of doing this is by adding another sodium ion to the products:

$$Na + 2H^+ \rightarrow \mathbf{2}Na^+ + H_2$$

Now check that the number of atoms still balances:
The Hs balance, but there are 2Nas on the right-hand side, and only one on the left.
So put a 2 in front of the left-hand side Na:

$$\mathbf{2}Na + 2H^+ \rightarrow 2Na^+ + H_2$$

The atoms **and** charges on each side balance, so that's your final answer.

Chemical Equations Sometimes Include **State Symbols**

State symbols show the **physical state** that a substance is in.
The state symbols you need to know about are in the box below:

(l) — liquid (g) — gas (s) — solid (aq) — aqueous (dissolved in water)

So the balanced equation for the reaction between hydrochloric acid and magnesium, including state symbols is: $Mg_{(s)} + 2HCl_{(aq)} \rightarrow MgCl_{2\,(aq)} + H_{2\,(g)}$.

Hold one ear and stare at something still — it'll help you balance...

1) Write a balanced symbol equation for the combustion of methane (CH_4) in oxygen.
   Step 1 has been done for you.
   Step 1: Methane + oxygen → carbon dioxide + water
2) Write balanced symbol equations for the following reactions.
   a) The complete combustion of ethanol (C_2H_5OH) in oxygen (O_2) to give carbon dioxide (CO_2) and water (H_2O).
   b) The reaction of calcium hydroxide ($Ca(OH)_2$) with hydrochloric acid (HCl) to give calcium chloride ($CaCl_2$) and water (H_2O).
3) Balance the following ionic equation: $Cl_2 + Fe^{2+} \rightarrow Cl^- + Fe^{3+}$.
   Include state symbols given that Cl_2 is a gas and everything else is aqueous.

Group 2

*Trend in **Reactivity** Down the Group*

During their reactions, Group 2 metals **donate** their **two outer electrons** to another atom. The reactivity of Group 2 metals depends on how **easily** the outer electrons can be donated. The **easier** the electrons can be donated, the **more reactive** the metal will be. You will find that:

Reactivity **increases** as you go **down** Group 2.

To see why, think about the factors that affect how strongly an electron is held by the nucleus:

1) The first is the **positive nuclear charge** — how **positive** the nucleus is. A **greater** nuclear charge provides a **stronger** force of attraction between the nucleus and electrons, and makes it more difficult for the atom to donate its outer electrons. As you go down the group, the nuclear charge **increases** as more **protons** are added to the nucleus, so if this was the **only** factor, reactivity would decrease down Group 2. But that **isn't** the case.
2) The second factor is that in **larger atoms**, the outer electrons are **further away** from the nucleus. The electrostatic attraction **decreases** in strength with **distance** from the source.
3) The third factor is **electron shielding**. As the atoms in Group 2 get **larger**, the number of **full electron shells** round the nucleus **increases**. These negative charges **shield** the two outer electrons from the attraction of the positive nucleus.

The increase in the **distance** between the outer electrons and the nucleus, and the increased **shielding** as you go down the Group, far **outweigh** the increase in nuclear charge.

4) You may have noticed that these are the **same factors** that affect the **ionisation enthalpy** (page 6). This is because both the reactivity of Group 2 and ionisation are to do with **removing electrons**.

*Trend in the **Melting Points** of Group 2 Metals*

You can see from the table that:

As you go **down** Group 2, **melting point decreases**.

Magnesium doesn't fit in with the general trend. It behaves a bit oddly because it has a slightly different structure to the other Group 2 metals.

	Melting Point (°C)
Beryllium (Be)	1278
Magnesium (Mg)	651
Calcium (Ca)	839
Strontium (Sr)	769
Barium (Ba)	727

This is also due to the increase in **electron shielding** as you go down the group.

Group 2 metals, like all other metals, are held together in a lattice structure by **metallic bonds** (page 16).

The **strength** of the metallic bonds depend on how strong the **attraction** is between the positive ions and the free electrons. The **more shielded** the positive nuclei are, the **weaker** the attraction will be, and so the **less energy** will be required to break the bond and melt the metal.

I love the Group 2 Metals — they're really trendy...

1) The following are descriptions of the reactions of Be and Ca with cold water. Use them to predict the reactions of Mg and Sr.
   - Beryllium will not react with cold water at all.
   - Calcium reacts steadily with cold water to produce hydrogen gas and calcium hydroxide.
2) Predict, with reasoning, the trend in boiling points of the Group 2 metals.

Group 7

Some **General Properties** of Group 7 Elements

Group 7 elements all have **7 electrons** in their outer shell. As a result these elements either:

1) Form **ionic compounds** by gaining an **extra electron** or,
2) **share** a pair of electrons and form a **covalent bond**.

In their elemental state, the halogens bond **covalently**, forming diatomic molecules (two atoms joined with a single covalent bond). In each case the atoms share an electron pair.

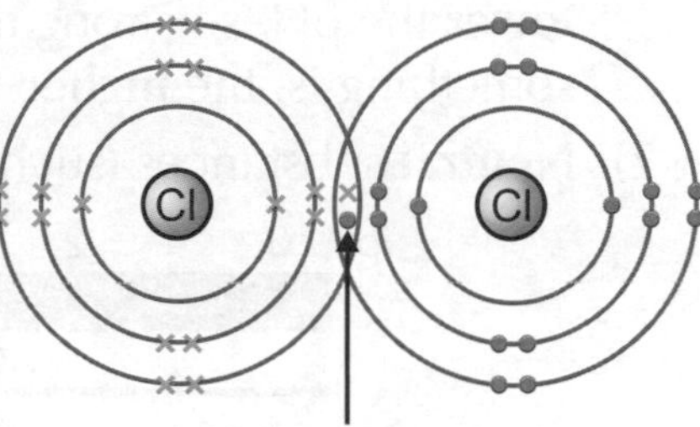

The halogen elements all have **coloured vapours**:

- **Chlorine** (Cl_2) is a **yellow/green gas** at room temperature.
- **Bromine** (Br_2) is a **brown liquid** at room temperature.
- **Iodine** (I_2) is a **grey solid** at room temperature (and sublimes to produce a **purple** vapour).

As you go down the group, the **melting points** and the **boiling points** of the elements **increase**. This is because the **strength** of the weak intermolecular bonds **between** molecules **increases** as the number of **electrons** in the molecules increases (see page 9).

Trend in **Reactivity** Down Group 7

During their reactions, Group 7 elements accept an **extra electron** from another atom. The reactivity of Group 7 elements depends on how **strongly** the nucleus can attract electrons. The **stronger** the attraction, the **more reactive** the element will be.

Reactivity **decreases** as you go **down** Group 7.

1) As with the Group 2 elements, **nuclear charge increases** as you go down the group. A greater nuclear charge will attract the extra electron required to fill the outer shell more strongly. This works to increase the reactivity of the elements as you go down the group.
2) However, as the atoms get bigger, the **extra shells** of electrons **shield** the nuclear charge more effectively. So the nucleus is **less able** to attract the extra electron the atom wants.

In Group 7 this **shielding** outweighs the effect of increasing nuclear charge. The elements at the **top** of the group are best able to attract an extra electron, and are more **reactive**.

Group 7 Reactivity and **Displacement Reactions**

You can show the relative reactivity of the Group 7 elements using **displacement reactions**. If you mix a **halogen** with a solution containing **halide ions**, a **more reactive** halogen will **displace** a **less reactive** halide ion (one below it in the group) from solution. For example:

Fluorine is more reactive than chlorine.

$$F_{2\,(aq)} + 2Cl^-_{(aq)} \rightarrow Cl_{2\,(aq)} + 2F^-_{(aq)}$$

The chloride ions have been **displaced** from the solution.

I met a friend for coffee today — I said 'Hallo, Jen'...

1) Predict, with reasoning, what would happen if you mixed the following halogens and halide solutions.
   a) Cl_2 and Br^-, b) I_2 and Cl^-, c) I_2 and Br^-, d) Cl_2 and I^-.
2) Draw a diagram to show the bonding between atoms in a fluorine molecule.

Acids and Bases

The pH Scale

1) The **pH** scale goes from **0** to **14** and measures how **acidic** or **basic** something is. **Acids** have a pH **less** than 7, while **bases** have a pH **greater** than 7. The **more acidic** something is, the **lower** the pH, so strong acids have a pH of between **0** and **1**. By contrast, the more **basic** something is, the **higher** its pH will be. **Strong bases** have a pH of **14**.
2) **Neutral** substances (such as water) have a pH of **7**. They are neither acidic nor basic.

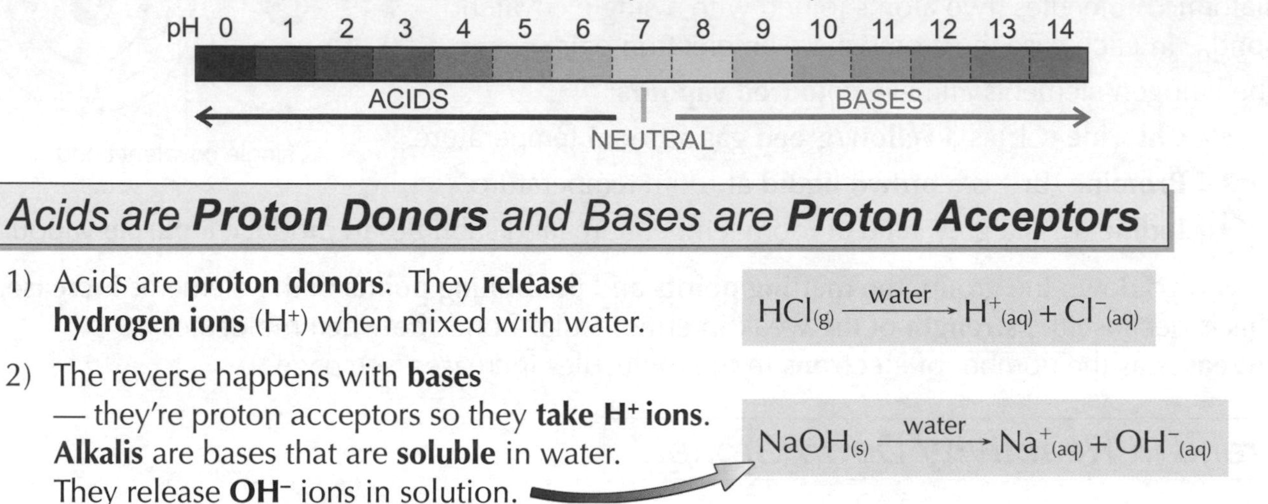

Acids are Proton Donors and Bases are Proton Acceptors

1) Acids are **proton donors**. They **release hydrogen ions** (H^+) when mixed with water.

$$HCl_{(g)} \xrightarrow{\text{water}} H^+_{(aq)} + Cl^-_{(aq)}$$

2) The reverse happens with **bases** — they're proton acceptors so they **take H^+ ions**. **Alkalis** are bases that are **soluble** in water. They release OH^- ions in solution.

$$NaOH_{(s)} \xrightarrow{\text{water}} Na^+_{(aq)} + OH^-_{(aq)}$$

3) When an acid reacts with an alkali, a **salt** and **water** are formed — this is called a **neutralisation** reaction.

acid + alkali → salt + water

You can show neutralisation just in terms of H^+ and OH^- ions. The hydrogen ions (H^+) from the acid will react with hydroxide ions (OH^-) from the base to produce water.

$$H^+_{(aq)} + OH^-_{(aq)} \rightarrow H_2O_{(l)}$$

EXAMPLE: Write a balanced equation for the reaction between hydrochloric acid (HCl) and sodium hydroxide (NaOH).

This reaction is a **neutralisation reaction** — a hydrogen ion from HCl combines with a hydroxide ion from NaOH to form water. The remaining ions combine to form the salt:

$$HCl_{(aq)} + NaOH_{(aq)} \rightarrow H_2O_{(l)} + NaCl_{(aq)}$$

Some Common Acids and Bases

Acid	Formula
Hydrochloric acid	HCl
Sulfuric acid	H_2SO_4
Nitric acid	HNO_3
Ethanoic acid	CH_3COOH

Base	Formula
Sodium hydroxide	$NaOH$
Potassium hydroxide	KOH
Ammonia	NH_3

Siobhan always tells the truth, but Alka lies...

1) Write a balanced equation for the reaction between nitric acid and potassium hydroxide.
2) Write equations to show what happens when the following substances are mixed with water:
a) sulfuric acid, b) potassium hydroxide, c) nitric acid.

Organic Molecules

There are Lots of **Families** of Compounds in Organic Chemistry

Organic Chemistry is the study of organic compounds — these are just substances that contain **carbon**. Carbon compounds can be split up into different **groups** which have similar **properties** and **react** in similar ways. These groups are called **homologous series**. All the compounds in a homologous series contain the same **functional group** — a certain group of atoms that is responsible for the **properties** of the molecule. Here are some **common homologous series**:

HOMOLOGOUS SERIES	FUNCTIONAL GROUP	EXAMPLE
alkanes	-C–C-	propane — $CH_3CH_2CH_3$
alkenes	-C=C-	propene — $CH_3CH{=}CH_2$
alcohols	-OH	ethanol — CH_3CH_2OH
aldehydes	C(=O)H	ethanal — CH_3CHO
ketones	C=O	propanone — CH_3COCH_3
carboxylic acids	-COOH	ethanoic acid — CH_3COOH

There are Different Ways of **Representing** a Molecule's **Structure**

Chemists have a few different ways of representing an organic molecule's **formula**.
Here are a few ways that you'll need to be able to interpret:

FORMULA	WHAT IT SHOWS YOU	FORMULA FOR BUTANOL (an alcohol)
General formula	This describes **any member** in a homologous series. The number of carbons is represented by 'n' and the number of hydrogens in terms of 'n'.	$C_nH_{2n+1}OH$ (this is true for all alcohols.)
Molecular formula	This shows the number of **atoms** of each **element** in a molecule.	$C_4H_{10}O$
Structural formula	This shows the molecule **carbon by carbon**, with all attached hydrogens and functional groups.	$CH_3CH_2CH_2CH_2OH$
Skeletal formula	The **bonds** of the carbon skeleton are drawn, with any **functional groups**. The carbon atoms and attached hydrogens aren't shown.	(zigzag chain)–OH
Displayed formula	All the **atoms** and **bonds** are drawn to show how the molecule is arranged.	H–C(H)(H)–C(H)(H)–C(H)(H)–C(H)(H)–O–H

Organic Chemistry — no pesticides, no added sugars, no flavourings...

1) Draw the skeletal and displayed formulae for the molecule with the structural formula $CH_3CHOHCH_2CH_3$.
2) What is the molecular formula of the compound with the structural formula CH_3CH_2COOH?

Alkanes

***Structure** and **Bonding** in Alkanes*

1) Alkanes are **hydrocarbons** — they **only** contain hydrogen and carbon atoms.
2) Alkanes contain **two types** of bond. All of the **carbon-carbon** bonds are **single covalent bonds**. All the other bonds are **carbon-hydrogen covalent bonds** (which are always single).
3) All of the available bonds have been formed, so we call alkanes **saturated** molecules.
4) **Carbon** always forms **four** covalent bonds and **hydrogen** makes **one** covalent bond.
5) The diagrams below show the structures of the first four **straight-chain alkanes**: methane, ethane, propane and butane.

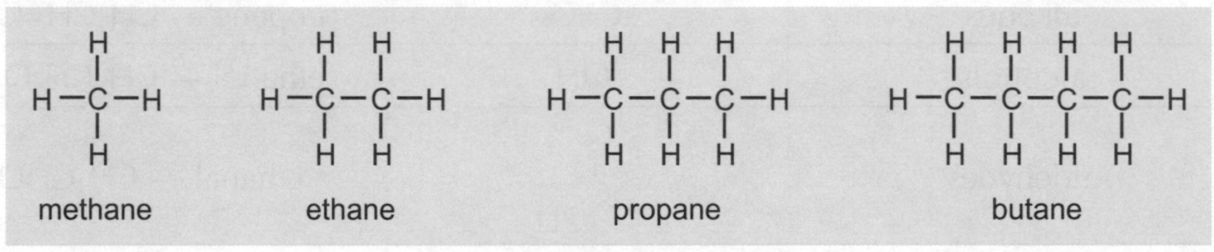

It is important to realise that these structures are only **2D representations** of the **3D molecules**. The molecules are not rigid. There is **free rotation** around a carbon-carbon single bond. This means that the carbon chains are quite **flexible** and gives the molecules the ability to **change shape**, particularly as the chain length increases.

***Properties** of Alkanes*

The **bonds** in alkanes are very **strong** and it requires a **lot of energy** to break them. This can be used to explain some of their **properties**:

1) They are very **unreactive**.
2) They are **not** able to form **polymers**.
3) They **burn cleanly**, tending to undergo **complete combustion** to form **carbon dioxide** and **water** (see page 28). The flame is usually a faint blue colour.
   For example, the combustion of ethane:
   $2C_2H_{6(g)} + 7O_{2(g)} \rightarrow 4CO_{2(g)} + 6H_2O_{(g)}$ (+ energy)

Also:

4) **Boiling point increases** as the **length** of the carbon chain increases.
5) **Viscosity** (resistance to flow) **increases** as chain length increases.
6) **Volatility** (ease of evaporation) **decreases** as chain length increases.

These last three properties are explained by the fact that the **attractive forces** between molecules get stronger as the chain length **increases** (page 9).

Alkanes are like the weather in the UK — completely saturated...

1) Draw out the structures of the next two alkanes, pentane (C_5H_{12}) and hexane (C_6H_{14}).
2) a) Write out the molecular formulae of the first four alkanes.
   b) We can work out a general formula for the alkanes of the form $C_nH_?$, where n is the number of carbon atoms. Work out, in terms of n, what should be in place of the ?.
3) Write a balanced equation for the complete combustion of propane in oxygen.

Alkenes

Structure and Bonding in Alkenes

1) **Alkenes** are similar to alkanes in that they are also **hydrocarbons**. The difference is in the presence of a **carbon-carbon double covalent bond** (C=C) somewhere in the carbon chain.
2) This means **not** all possible single bonds have been made — these molecules are **unsaturated**.
3) As in all compounds the carbon atoms must have **four** bonds, and hydrogen only **one**.
4) The structures of the first three alkenes (ethene, propene and butene) are shown below:

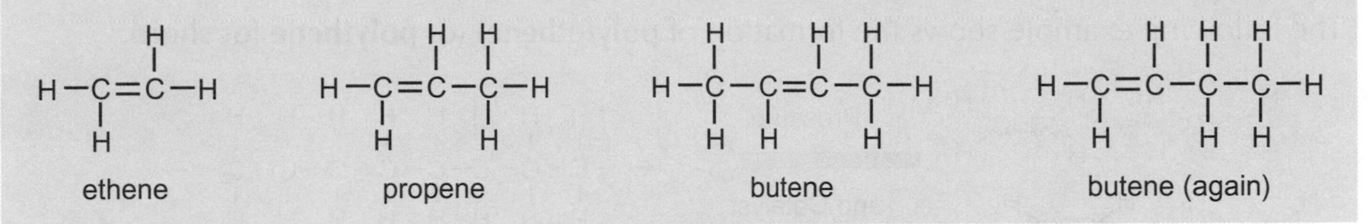

As you can see from butene, the presence of the C=C bond means that most alkenes have **more than one** possible structure. The C=C bond can be in various different **positions** along the chain.

Molecules with the same **molecular formula** but different **structures** are called **isomers**.

The C=C bond does not allow the same **free rotation** and flexibility around itself as a C–C bond. It is a **rigid** bond. But the rest of the carbon chain is the same as in an alkane molecule, so rotation is allowed around the **single** bonds.

Properties of Alkenes

The presence of the C=C bond dictates the chemical properties of alkenes.

1) They are **reactive** compounds, undergoing many different types of chemical reaction.
2) They are used extensively to form **polymers**, e.g. poly(ethene) (see next page).
3) They **do not** burn **cleanly**, giving very **yellow** flames and lots of **soot**.

And as for alkanes, when you increase the **chain length** of an alkene:

4) The **boiling point** increases.
5) The **viscosity** increases.
6) The **volatility** decreases.

You can **test** for whether a compound is an alkene by adding it to **bromine water**. Alkenes **decolourise** bromine water, turning it from **orange** to **colourless**.

bromine water + alkene → SHAKE → solution goes colourless

'Sleeping Butene' — coming soon to a cinema near you...

1) Draw out a structure for the next alkene: pentene (C_5H_{10}).
2) Draw out two alternative structures for hexene (C_6H_{12}).
3) Work out the general formula for the alkenes of the form $C_nH_?$.

Polymerisation

Alkenes Can Form **Polymers**

The presence of the double bond in alkene molecules means that they are capable of forming **polymers**. A polymer is a long, chain-like molecule built up from lots of **repeating units**. In this case the repeating units, called **monomers**, are alkene molecules.

Under the right conditions (these depend on the alkene and the desired properties of the polymer), many small alkenes (like ethene and propene) will **open** up their double bonds and **link together** to form these long chain polymers.

The following example shows the formation of **poly(ethene)** (or **polythene** for short):

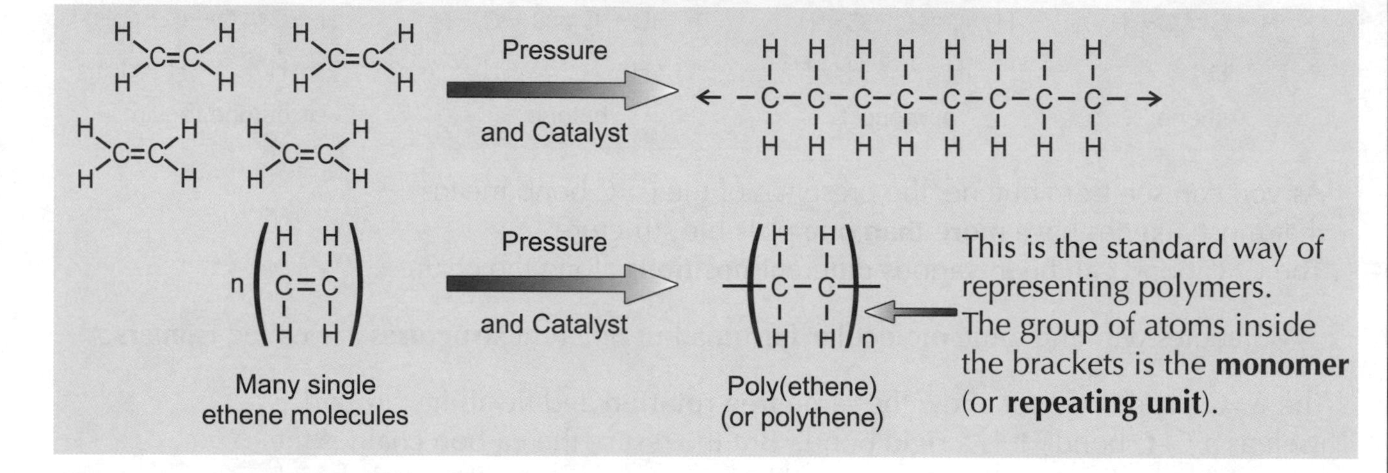

Other Small Alkenes do a **Similar Thing**

1) **Propene** polymerises to form **polypropene**.

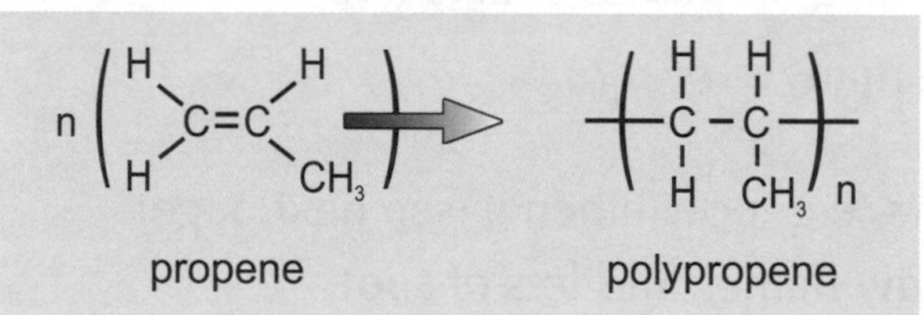

2) **Styrene**, which has a **benzene** ring in it, polymerises to form **polystyrene**.
(Benzene is just a ring of six carbon atoms in which the bonding electrons are shared between all six carbons.)

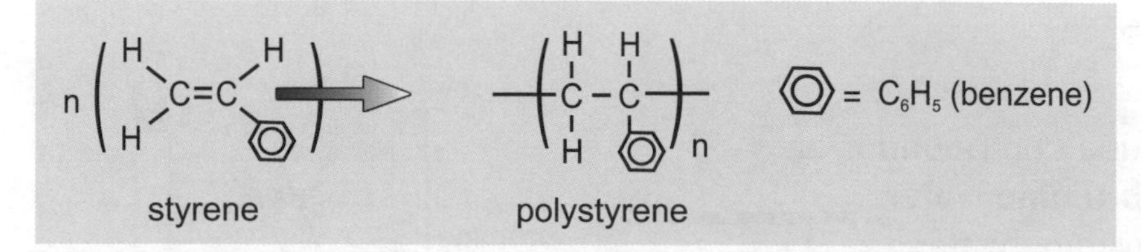

I'd go on and on about how great polymers are, but it'd get repetitive...

1) What property of alkenes allows them to form polymers?
2) Using the standard way of representing polymers, shown above, draw the polymers formed by the following alkenes:
   a) chloroethene: $H_2C{=}CHCl$
   b) this isomer of butene: $H_3C(H)C{=}C(H)CH_3$

Alcohols

Alcohols Contain an -OH Group

The **alcohols** are a group of compounds that all contain an -OH group (an oxygen atom covalently bonded to a hydrogen atom).

The first three alcohols are called **methanol**, **ethanol** and **propanol**. Their structures are:

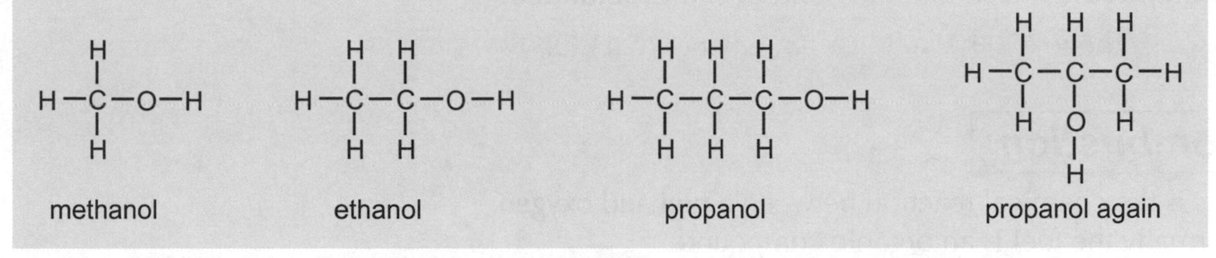

For many alcohols, the -OH group can be put in different **positions** along the chain, so they are able to form **isomers** — just like in the example with propanol above.

Alcohols can be called **primary**, **secondary** or **tertiary**. The type of alcohol depends on what other groups surround the carbon atom that the -OH group is attached to.

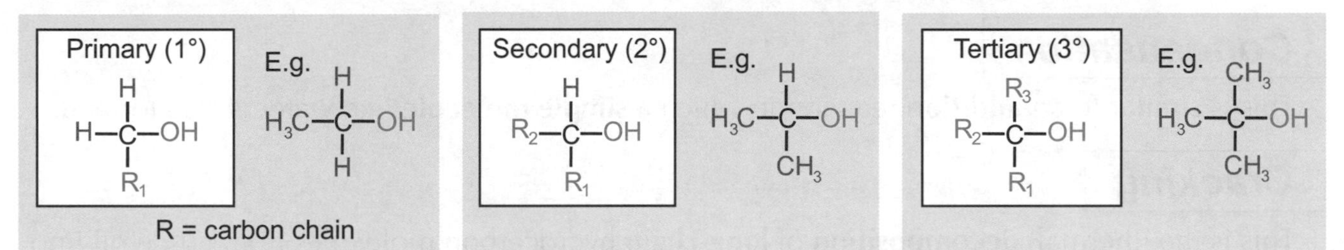

The Properties of Alcohols

Oxygen is an **electronegative** element (see page 10), so it draws the bonding electrons towards itself in the C–OH bond, meaning that alcohols are normally **polar** molecules.

$$\overset{\delta+}{R_1}-\overset{\delta-}{O}-\overset{\delta+}{H}$$

The electronegative oxygen also draws electrons away from the **hydrogen atom** in the -OH group, giving the hydrogen atom a **slightly positive charge**. This charge attracts the **lone pairs** of **electrons** on oxygen atoms in other nearby alcohol molecules, which forms a **hydrogen bond** (page 10).

Hydrogen bonds have a big effect on the **properties** of alcohols.

1) Alcohols are **soluble** in water.
2) Alcohols have **high boiling** and **melting** points compared to alkanes or alkenes of a similar size. This is because hydrogen bonds are the **strongest** type of intermolecular bond, so they need lots of energy to break.

Alcohols — always wine-ing about their rum luck. I cava beer it...

1) Draw two different isomers of butanol, C_4H_9OH.
2) Work out the general formula of alcohols, using the form $C_nH_?OH$.
3) Predict, with reasoning, whether ethane or ethanol will have a higher melting point.

Reaction Types

There are lots of types of **chemical reaction**. You will need to know all of them quite well.
These pages give you types, explanations and examples (in alphabetical order).

Addition

This is a reaction in which atoms are **added** to an **unsaturated** bond so that the bond becomes **saturated**.

e.g. ethene (C_2H_4) + H_2O → ethanol (C_2H_5OH)

Combustion

This is the chemical reaction between a **fuel** and **oxygen**.
Normally the fuel is an organic compound and the products are **carbon dioxide** and **water** — this is **complete combustion.**

e.g. $C_3H_8 + 5O_2 \rightarrow 3CO_2 + 4H_2O$

Without enough oxygen, **incomplete combustion** occurs, producing poisonous carbon monoxide.

e.g. $C_3H_8 + 3½O_2 \rightarrow 3CO + 4H_2O$

Condensation

This is similar to an **addition** reaction in which a **simple molecule** like water is also formed.

Cracking

This is the (thermal) **decomposition** of **long-chain** hydrocarbon molecules from crude oil into **shorter-chain** alkanes and alkenes. This requires **high temperatures** and **pressures** and a **catalyst** (usually aluminium oxide), and makes hydrocarbons that are more useful.

e.g. decane ($C_{10}H_{22}$) → octane (C_8H_{18}) + ethene (C_2H_4)

Dehydration

This is the removal of **water** from a compound by **heating**.
In organic molecules it usually results in the formation of a **C=C** bond.

e.g. ethanol (C_2H_5OH) → ethene (C_2H_4) + H_2O

Displacement

This is a reaction where one element **displaces** another, **less reactive**, element from a compound. This usually takes place between **metals**, but also with **halogens**.

e.g. $2Al_{(s)} + Fe_2O_{3(s)} \rightarrow Al_2O_{3(s)} + 2Fe_{(s)}$ (the Thermite Reaction)

Disproportionation

This is a rare type of chemical reaction where an **element** in a reactant is **oxidised** and **reduced** at the same time. **Chlorine** can undergo disproportionation reactions.

e.g. $Cl_{2(aq)} + H_2O_{(l)} \rightarrow HOCl_{(aq)} + HCl_{(aq)}$

Chloric(I) acid Hydrochloric acid

The chlorine has been: **oxidised** **reduced**

Reaction Types

Electrolysis

This is a process that uses **electricity** to **break down** a compound. The reactant or reactants must be in the **liquid** state — either **molten** or in **solution**. The particles have to be able to move.
An example is the electrolysis of bauxite to obtain pure aluminium.

Elimination

This is just the **removal** of a **small molecule** from a larger molecule.
Usually H_2O or H_2 is removed (and not replaced by anything else).

e.g. propanol ($CH_3CHOHCH_3$) + sulfuric acid catalyst → propene ($CH_2=CHCH_3$) + water

Endothermic

Any chemical reaction that **takes in** heat energy. This means that the **reactants** will have **less energy** than the **products**. The **enthalpy change** of reaction, ΔH (see page 41), is **positive**.

Exothermic

Any chemical reaction that **gives out** heat energy. This happens because the **products** have **less energy** than the **reactants**. The **enthalpy change** of reaction, ΔH, is **negative**.

Hydrogenation

This is the **addition** of a molecule of **hydrogen** (H_2) across a **C=C** bond.
One atom attaches to each carbon.

e.g. ethene (C_2H_4) + H_2 → ethane (C_2H_6)

Neutralisation

This is the reaction between a **basic compound** and an **acid**. The products always include the **salt** of the acid, **water** and other products dependent on the acid and base.

e.g. $2KOH_{(aq)} + H_2SO_{4(aq)} \rightarrow K_2SO_{4(aq)} + 2H_2O_{(l)}$

$Na_2CO_{3(aq)} + 2HCl_{(aq)} \rightarrow 2NaCl_{(aq)} + CO_{2(g)} + H_2O_{(l)}$

Oxidation

There are two possible definitions for this — the best is the **loss of electrons**.
Another useful one is the **gain of oxygen**. It is the opposite of reduction.

Precipitation

A precipitate is a **solid** that is formed in a **solution** by a chemical reaction or by a change in temperature affecting solubility. Precipitates are **insoluble** in the solvent.
A precipitation reaction is simply any reaction that **produces a precipitate**.

Reaction Types

Radical (Chain) Reactions

Reactions involving radicals — an atom or compound with an **unpaired electron**. Often, one of the **products** of the reaction is also a radical which can perform further reactions. This makes the process a **chain reaction**.

Redox

This is the name for a reaction that involves both **reduction** and **oxidation** processes. It is usually used to describe reactions that just involve **electron transfer**.

e.g. $Fe_{(s)} + Cu^{2+}_{(aq)} \rightarrow Fe^{2+}_{(aq)} + Cu_{(s)}$

reduction ($Cu^{2+} \rightarrow Cu$)
oxidation ($Fe \rightarrow Fe^{2+}$)

Reduction

There are two possible definitions for this — the best is the **gain of electrons**. The other useful one is the loss of oxygen. Important point: oxidation and reduction **ALWAYS** happen **together** — it is impossible to have one without the other.

Reversible

This is the name given to any chemical reaction that can go **forwards** and **backwards** at the **same time**. That means that the reactants will form the products, but that the products will also react (or decompose) to give the reactants.

e.g. $N_{2(g)} + 3H_{2(g)} \rightleftharpoons 2NH_{3(g)}$

Substitution

This is simply a reaction in which an atom (or group of atoms) in a molecule is **swapped** for a different atom (or group of atoms).

Thermal Decomposition

This is where one compound **breaks down**, under **heating**, into two or more simpler compounds. A classic example is the breakdown of any carbonate compound,

e.g. $CaCO_3 + \text{heat} \rightarrow CaO + CO_2$

Cracking of hydrocarbons is also an example.

I'm in the middle of a chain reaction...

1) Write down all the different types of reaction that each of the following could be classed as.
   a) burning ethanol
   b) iron + copper sulfate → iron sulfate + copper
   c) hydrochloric acid + sodium hydroxide → sodium chloride + water + heat
   d) propene (C_3H_6) + H_2 → propane (C_3H_8)

Reaction Rates

Measuring the Rate of a Reaction

The **rate** of reaction is just a measure of how **fast** a particular reaction is going.

You need to know some of the ways that you can follow the rates of different reactions. They're all about measuring how fast the **reactants** are being **used up**, or measuring how fast the **products** of the reaction are **forming**.

There are lots of ways of measuring the rate of a reaction:

1) You can measure the **change in mass** that occurs during a reaction where gas is released as one of the products.

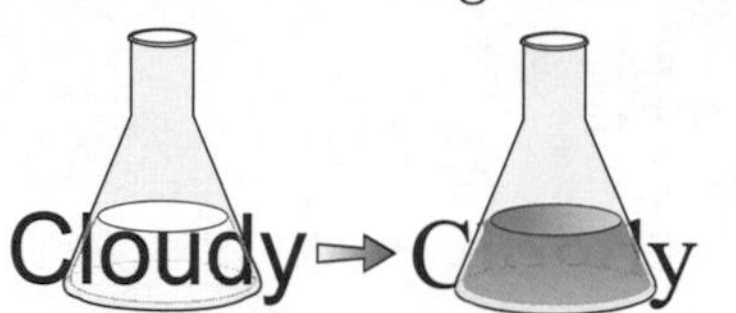

2) You can follow the **colour change** of a reaction. This includes precipitation reactions, where the solution turns cloudy as more of the product is made.
3) You can measure changes in **temperature** or **pH** that occur during the reaction.
4) You can measure the **volume of gas** produced during a reaction.

EXAMPLE: Measuring the rate of reaction between hydrochloric acid and magnesium metal.

magnesium + hydrochloric acid → magnesium chloride + hydrogen

- Use a **gas syringe** to collect the hydrogen gas that is given off during the reaction.
- Use a **stopwatch** to **time** the reaction.
- At **timed intervals**, say every 30 seconds, **record** how much hydrogen gas has been produced.

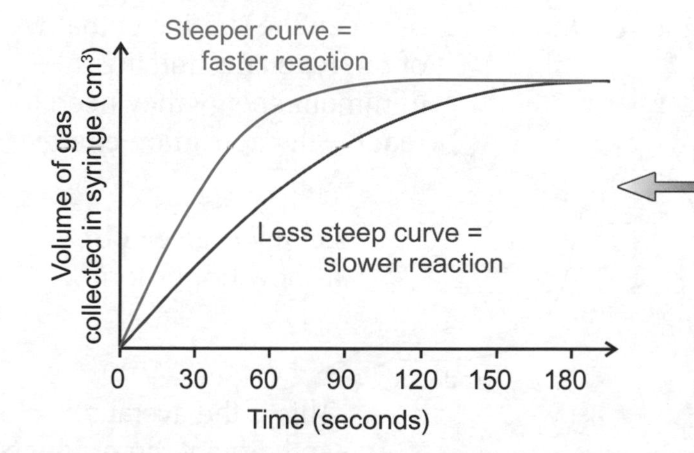

Plotting graphs lets you compare rates of reactions.

(Another way to measure the rate of this reaction would be to measure the decrease in **mass** as hydrogen gas is lost from the reaction container.)

My rate of chocolate biscuit consumption is worryingly high...

1) Describe how you could measure the rates of the following reactions:
   a) The endothermic reaction between citric acid and sodium bicarbonate to give carbon dioxide, water and a sodium salt.
   b) The precipitation reaction between sodium thiosulfate and hydrochloric acid to form a sulfur precipitate, sulfur dioxide gas, sodium chloride and water.
   c) The reaction between solid calcium carbonate and hydrochloric acid to produce calcium chloride and carbon dioxide gas.

Collision Theory

Particles Need to **Collide** in Order to **React**

Reaction rates are explained by **collision theory**. It's based on the idea that particles in liquids and gases are always **moving around** and **colliding** with each other. Not every **collision** results in the particles **reacting**. The following **conditions** need to be right:

- The particles need to collide in the **right direction**. They need to be **facing** each other the right way.
- The particles need to collide with at least a certain **minimum** amount of **energy**.

Collision theory states that the **more collisions** there are, and the **more energy** these collisions have, the **more likely** particles are to react.

Particles Need Enough **Energy** to React

1) The **minimum amount of kinetic** (movement) **energy** particles need to react is known as the **activation energy**. This energy is used to **break the bonds** to start the reaction.
2) Reactions with **low activation energies** often happen **pretty easily**, but reactions with **high activation energies** don't — you have to give the particles **extra energy** (e.g. by **heating** them).

To make this a bit clearer, here's an **enthalpy profile diagram** — enthalpy is just a fancy word for energy. These diagrams show how the **enthalpy** of the reacting particles changes over the course of a reaction (see page 41 for more on enthalpy changes).

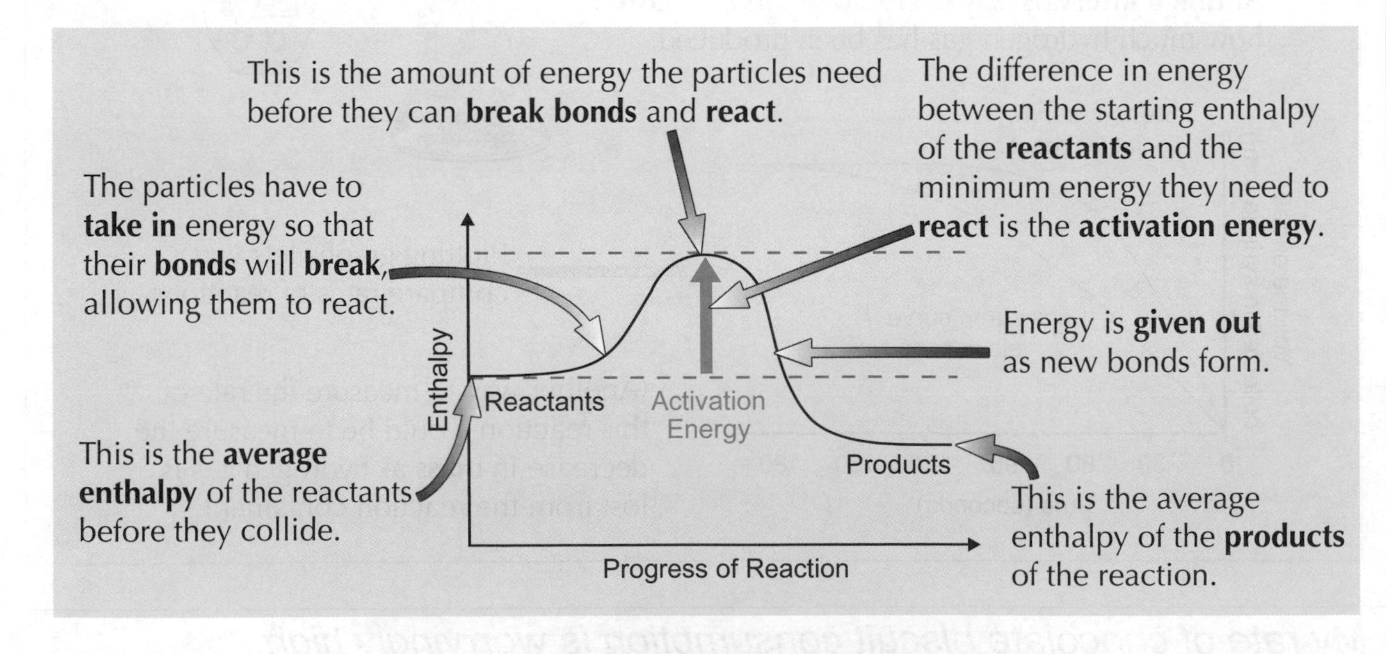

Toast and a large cup of tea — that's my morning activation energy...

1) Two particles in a reaction vessel collide but don't react. Give two reasons why the reaction may not have happened.
2) What is the activation energy of a reaction?
3) Draw an enthalpy profile diagram for a reaction. On your diagram, label the reactants, products and activation energy.

Reaction Rates and Catalysts

Changing the Rate of Reaction

The rate of reaction depends on **how often** particles **collide** (see page 32) and how **likely** the collisions are to be **successful**.
More frequent **successful** collisions mean a **faster** rate of reaction.

These factors all **increase** the rate of reaction:

1) **Increasing temperature** — the particles tend to have more kinetic energy. This means that they move around faster, and so are more likely to collide with each other **and** have enough energy to react.
2) **Increasing concentration (or pressure in gases)** — this means that the particles of reactant will be closer together, so they will be more likely to collide.
3) **Increasing the surface area of a solid reactant** — this increases the number of particles of the solid reactant able to come into contact with other reactants.

EXAMPLE: Predict whether magnesium dust or magnesium ribbon will react faster with hydrochloric acid.

Magnesium dust has a larger **surface area** than magnesium ribbon.
Increasing the surface area of a solid reactant increases the rate of reaction, so **magnesium dust** will react faster than magnesium ribbon.

Catalysts Speed Up Reactions

You met **activation energy** on the last page — it's just the **minimum** amount of energy needed for a reaction to happen.
A **catalyst increases** the **rate** of a reaction by **lowering** its activation energy.

A **catalyst** is any substance which changes the **rate** of a reaction, without being **changed** or **used up** itself.

Catalysts are also very **specific** — different reactions will only be sped up by **certain catalysts**.
There are loads of advantages to using catalysts:

1) Catalysts reduce the need for **high temperatures** and **pressures** in industrial reactions, like hydrocarbon cracking (see page 28) and ethanol production (see page 36). This makes these processes **cheaper** to run.
2) Using lower temperatures also means less **energy demand**, and so lower CO_2 emissions.

Tabbys are number one on my cat list...

1) Describe two things you could do to increase the rate of a reaction between aqueous species.
2) Why does increasing the pressure increase the rate of a reaction between gases?
3) What's a catalyst?
4) Give two advantages of using a catalyst in industrial reactions.

Reversible Reactions

Reversible Reactions Go **Both Ways**

In a reversible reaction, the **products** can react with each other and **change back** into the reactants.

Reactants	Products
$A + B \rightleftharpoons$	$C + D$

So there are actually two reactions happening at once: $A + B \rightarrow C + D$ and $C + D \rightarrow A + B$. This can affect the **yield** of a reaction, as some of the products will be converted **back** into reactants.

EXAMPLE: The industrial production of ethanol from ethene.

exothermic →

$$H_2C{=}CH_{2(g)} + H_2O_{(g)} \rightleftharpoons CH_3CH_2OH_{(g)}$$

← endothermic

Catalyst: H_3PO_4
Temperature: 300 °C
Pressure: 60 atm

Because the reaction is reversible you **don't** get a **high yield** — some of the ethanol **converts back** to ethene and water. But you can keep **removing** and **recycling** any ethene that you have left, so you can still end up with a good overall yield.

Reversible Reactions Reach an **Equilibrium**

If a reversible reaction is taking place in a **closed system** it will eventually reach a state of **equilibrium**.

A **closed system** is one where nothing can **get in** or **out**.

1) When a reaction **begins** there will be a **high concentration** of **reactants**, and a **low concentration** of **products** in the system. So the **forward** reaction will be **fast**, and the **reverse** reaction quite **slow**.
2) The concentration of **reactants** will gradually **decrease**, while the products build up. So the **forward** reaction will start to **slow down** while the **reverse** reaction **speeds up**.
3) After a while the forward reaction and the reverse reaction end up going at the **same rate**. From this point on the **concentration** of the **reactants** and **products won't change**.
4) This is called **dynamic equilibrium**. The forward and reverse reactions are **both still happening** — some reactant is being made into product, and some product is being made into reactant.
5) But since these processes are going at **exactly the same rate**, it seems as if nothing's happening.

Dynamic equilibrium — like walking up a down escalator...

1) Compare the rates of the forward and backward reactions of a reversible reaction at the following points:
   a) At the start of the reaction.
   b) At equilibrium.
2) What is dynamic equilibrium?

Le Chatelier's Principle

***Position** of Equilibrium*

The **position** of equilibrium tells you the amount of **reactants compared** to the amount of **products** that are present when the reaction reaches an **equilibrium**.

$$\text{Reactants} \quad \text{Products}$$
$$A + B \rightleftharpoons C + D$$

If the position of equilibrium lies on the **left-hand side**, there are **more reactants** than products in the reaction mixture.

If the position of equilibrium lies on the **right-hand side**, there are more **products** than reactants in the reaction mixture.

*Changing **Conditions** Changes the **Equilibrium Position***

Altering the conditions of a reversible reaction can **move** the position of equilibrium in one direction or the other. Careful control of the conditions can result in a higher yield (more of the products).

Look at the production of ethanol from ethene again as an example:

exothermic →

$$H_2C{=}CH_{2(g)} + H_2O_{(g)} \rightleftharpoons CH_3CH_2OH_{(g)}$$

← endothermic

1) If you increase the **pressure**, conditions will favour the forward reaction and **more ethanol** (CH_3CH_2OH) will be formed.
   This is because there are **more molecules** of gas on the **left-hand side** than on the right-hand side — two molecules of $H_2C{=}CH_2/H_2O$ react to form **only one** molecule of CH_3CH_2OH. This **reduces** the pressure.
2) Raising the **temperature** favours the **reverse** reaction. This is because it's **endothermic** (see page 41) and **absorbs** the extra heat energy, **lowering** the temperature.
3) **Removing ethanol** from the container as it forms will push the equilibrium to the **right** to try and make up for the change in concentration between the reactants and products.

These observations can be summarised by an important rule known as **Le Chatelier's Principle**:

A **reversible reaction** will move its **equilibrium position** to **resist** any **change** in the conditions.

Equilibrium reactions are so stubborn — always resisting change...

1) You are making ethanol from ethene and steam using the reaction shown above. What will happen to the yield of ethanol if you increase the amount of steam in the reaction mixture?
2) Ammonia is produced industrially using the following reversible reaction:
   $$N_{2(g)} + 3H_{2(g)} \rightleftharpoons 2NH_{3(g)}$$
   The forward reaction is exothermic and the backwards reaction is endothermic.
   How will the position of the equilibrium change if you:
   a) Increase the temperature of the reaction?
   b) Remove some ammonia from the reaction?

Equilibrium and Yield

Deciding on the **Best Conditions** to Use

Thanks to Le Chatelier's principle (see page 35) you might think it would be **easy** to work out the **optimum conditions** for a reversible reaction. But in real life it's not quite that simple.

For most reversible reactions that are used on an industrial scale there are other factors, such as **cost** and **time**, that need to be taken into account.

Have a look at the conditions used for the production of ethanol from ethene again:

$$H_2C{=}CH_{2(g)} + H_2O_{(g)} \rightleftharpoons CH_3CH_2OH_{(g)}$$

exothermic (forward) → ; ← endothermic (reverse)

Catalyst: H_3PO_4
Temperature: 300 °C
Pressure: 60 atm

Temperature:

1) **Lowering the temperature** would favour the forward reaction and so it should increase the **yield** of ethanol.
2) But lowering the temperature also means that fewer of the particles in the reaction mixture will have **enough energy** to react (see page 32). The particles will also be moving **more slowly**, so there will be **fewer collisions**. So lowering the temperature will **slow down** the **rate** of both the forward and reverse reactions.
3) A low temperature would make the forward reaction **too slow** to be useful. So a compromise temperature of **300 °C** is used.

Pressure:

1) **Increasing the pressure** would favour the forward reaction and increase the **rate** of reaction (as the particles will be **closer together** so will collide **more frequently**). This would increase the **yield** of ethanol.
2) But producing high pressures uses a lot of **energy** and **costs** a lot of money. You'd need some pretty strong equipment to stand up to the high pressures too — and that would be expensive to buy and maintain.
3) To make the reaction economic, a moderately high pressure of **60 atm** is used.

Concentration:

1) Ethanol is **removed** from the reaction vessel as it is produced.
2) This reduces the concentration of products so the equilibrium shifts to favour the **forwards reaction**. This **improves** the **yield** of ethanol.

Catalyst:

1) Using a solid **phosphoric acid(V)** catalyst **increases** the rate of **both** the forward and the backward reactions.
2) The catalyst has **no effect** on the **position** of the equilibrium — it just means the equilibrium is reached **faster** and the **temperature** and **pressure** at which the reaction can happen, at a reasonable rate, are **reduced.**

I should put a dodgy pun here, but I won't yield to the pressure...

1) Explain why the reaction above is not run industrially at a temperature of 40 °C.
2) Explain why the reaction above is not run industrially at a pressure of 500 atm.

The Mole

A Mole is a Number of Particles

If you had a sample of a substance, and you wanted to **count** the number of atoms that were in it, you'd have to use some very **big numbers**, and spend a very long time counting. So you need a **unit** to describe the **amount** of a substance that you have — that unit is the **mole**.

One mole of a substance contains $\mathbf{6.02 \times 10^{23}}$ particles.
6.02×10^{23} mol^{-1} is known as **Avogadro's constant**.

The particles can be **anything** — e.g. atoms or molecules (or even giraffes).
So 6.02×10^{23} atoms of **carbon** is 1 mole of carbon,
and 6.02×10^{23} molecules of CO_2 is 1 mole of CO_2.

Molar Mass is the Mass of One Mole

One mole of atoms or molecules has a **mass in grams** equal to the **relative formula mass** (A_r or M_r) of that substance.

For **carbon**, A_r = **12.0** so 1 mole of carbon weighs **12 g** and the **molar mass** is **12 g mol^{-1}**.
For **CO_2**, M_r = **44.0** so 1 mole of CO_2 weighs **44 g** and the **molar mass** of CO_2 is **44 g mol^{-1}**.
So, **12.0 g** of **carbon** and **44.0 g** of **CO_2** must contain the **same number of particles**.

You can use molar mass in calculations to work out how many moles of a substance you have.
Just use this formula:

$$\text{Number of moles} = \frac{\text{Mass of substance (g)}}{\text{Molar mass (g mol}^{-1})}$$

g mol^{-1} is the same as g/mol.

EXAMPLE: How many moles of sodium oxide are present in 24.8 g of Na_2O?

Molar mass of Na_2O = (2 × 23.0) + (1 × 16.0) = 62.0 g mol^{-1}
Number of moles of Na_2O = 24.8 g ÷ 62.0 g mol^{-1} = **0.400 moles**

You can **rearrange** the formula above and use it to work out the mass of a substance or its relative formula mass (see page 3). It can help to remember this triangle:

mass / moles | molar mass

EXAMPLE: What is the mass of 1.30 moles of magnesium oxide (MgO)?

Molar mass of MgO = (1 × 24.3) + (1 × 16.0) = 40.3 g mol^{-1}
Rearranging the formula, mass = moles × molar mass
So mass of MgO = 1.30 × 40.3 = **52.4 g** (3 s.f.)

Avocado's constant: how much I need to satisfy my guacamole craving...

1) Find the molar mass of sulfuric acid, given that 0.700 moles weighs 68.6 g.
2) How many moles of sodium chloride are present in 117 g of NaCl?
3) I have 54.0 g of water (H_2O) and 84.0 g of iron (Fe). Do I have more moles of water or of iron?

Determination of Formulae from Experiments

Empirical and *Molecular* Formulae

The **empirical formula** of a compound is the **simplest ratio** of the atoms of each element in the compound.

The **molecular formula** of a compound gives the **actual number** of atoms of each element in the compound.

For example, a compound with the molecular formula C_2H_6 has the empirical formula CH_3. The **ratio** of the atoms is one C to every three Hs.

Calculating *Empirical* Formulae

Often, the only way to find out the formula of a compound is through **experimentation** and **calculation**. You can calculate the formula of a compound from the **masses** of the **reactants**. Here is a simple set of rules to follow when calculating a formula:

1) Write the **mass** or **percentage mass** of each element.
2) Find the number of **moles** of each substance by dividing by the atomic or molecular mass.
3) Divide all answers by the **smallest** answer.
4) If required: multiply to make up to **whole numbers**.
5) Use the **ratio** of atoms to write the formula (this gives the empirical formula).

EXAMPLE: Find the formula of an oxide of aluminium formed from 9.00 g aluminium and 8.00 g oxygen.

1) First write down the mass of each substance:
   Al: 9.00 g　　O: 8.00 g
2) Divide the mass by the atomic masses to find the number of moles of each substance:
   Al: 9.00 ÷ 27.0 = 0.333 moles　　O: 8.00 ÷ 16.0 = 0.500 moles
3) Divide by the smallest number, which is 0.333:
   Al: 0.333 ÷ 0.333 = 1.00　　O: 0.5 ÷ 0.333 = 1.50
4) Multiply by 2 to give whole numbers:
   Al: 1.00 × 2 = 2　　O: 1.50 × 2 = 3
5) The ratio of Al : O is **2 : 3**.
   The empirical formula is Al_2O_3.

Roman empirical formula — 1 Caesar, 3 gladiators & 8 straight roads...

1) Find the empirical formulae of the following oxides:
   a) An oxide containing 12.9 g of lead to every 1.00 g of oxygen.
   b) An oxide containing 2.33 g of iron to every 1.00 g of oxygen.
   (Relative atomic mass values: Pb = 207.2, O = 16.0, Fe = 55.8)
2) Calculate the empirical formula of the carboxylic acid that is comprised of 4.30% hydrogen, 26.1% carbon and 69.6% oxygen.
   (Relative atomic mass values: H = 1.0, C = 12.0, O = 16.0)

Calculation of Molecular Formulae

*Use the **Relative Formula Mass** to Work Out the Molecular Formula*

To find the **molecular formula** from the **empirical formula**, you need to know the **relative formula mass** (see page 3) of the compound. This will usually be given to you in the question. Read through the example below and then try the questions.

EXAMPLE: Calculate the molecular formula of a hydrocarbon molecule if the compound contains 85.7% carbon and it's relative formula mass is 42.0.

First calculate the empirical formula:
In 100 g of the compound, there will be:
C: 85.7 g H: (100 g – 85.7 g) = 14.3 g
Number of moles of each compound:
C: 85.7 ÷ 12.0 = 7.14 H: 14.3 ÷ 1.0 = 14.3
Divide by the smallest number (7.14):
C: 7.14 ÷ 7.14 = 1 H: 14.3 ÷ 7.13 = 2
So the ratio of C:H is **1:2**.
The empirical formula is $\mathbf{CH_2}$.

Hydrocarbons only contain carbon and hydrogen, so any mass that isn't carbon will be hydrogen.

Calculate how many multiples of the empirical formula the molecular formula contains:
The empirical formula (CH_2) has a relative mass of 12.0 + 1.0 + 1.0 = 14.0.
The molecular formula has a relative mass of 42.0.
42.0 ÷ 14.0 = 3
To find the molecular formula, multiply each of the values in the empirical formula by 3:
C: 1 × 3 = 3 H: 2 × 3 = 6
The molecular formula is $\mathbf{C_3H_6}$.

The example above uses **percentage compositions** rather than the **mass** of each element in the compound. You can calculate the **percentage composition** yourself using the formula:

$$\text{percentage composition of element X} = \frac{\text{total mass of element X in compound}}{\text{total mass of compound}} \times 100\%$$

The percentage composition of my fridge is 80% cheese & 20% juice...

1) Calculate the molecular formula of a compound containing 52.2% carbon, 13.0% hydrogen and 34.8% oxygen if the relative formula mass of the compound is 46.0.
(Relative atomic mass values: C = 12.0, H = 1.0, O = 16.0)
2) Calculate the molecular formula of a hydrocarbon with a relative formula mass of 78.0 that contains 92.3% carbon.
(Relative atomic mass values: C = 12.0, H = 1.0)
3) Find the percentage composition of oxygen in each of the following compounds:
   a) Ethanol (C_2H_5OH).
   b) Nitric acid (HNO_3).
   c) Propanone (C_3H_6O).

Atom Economy

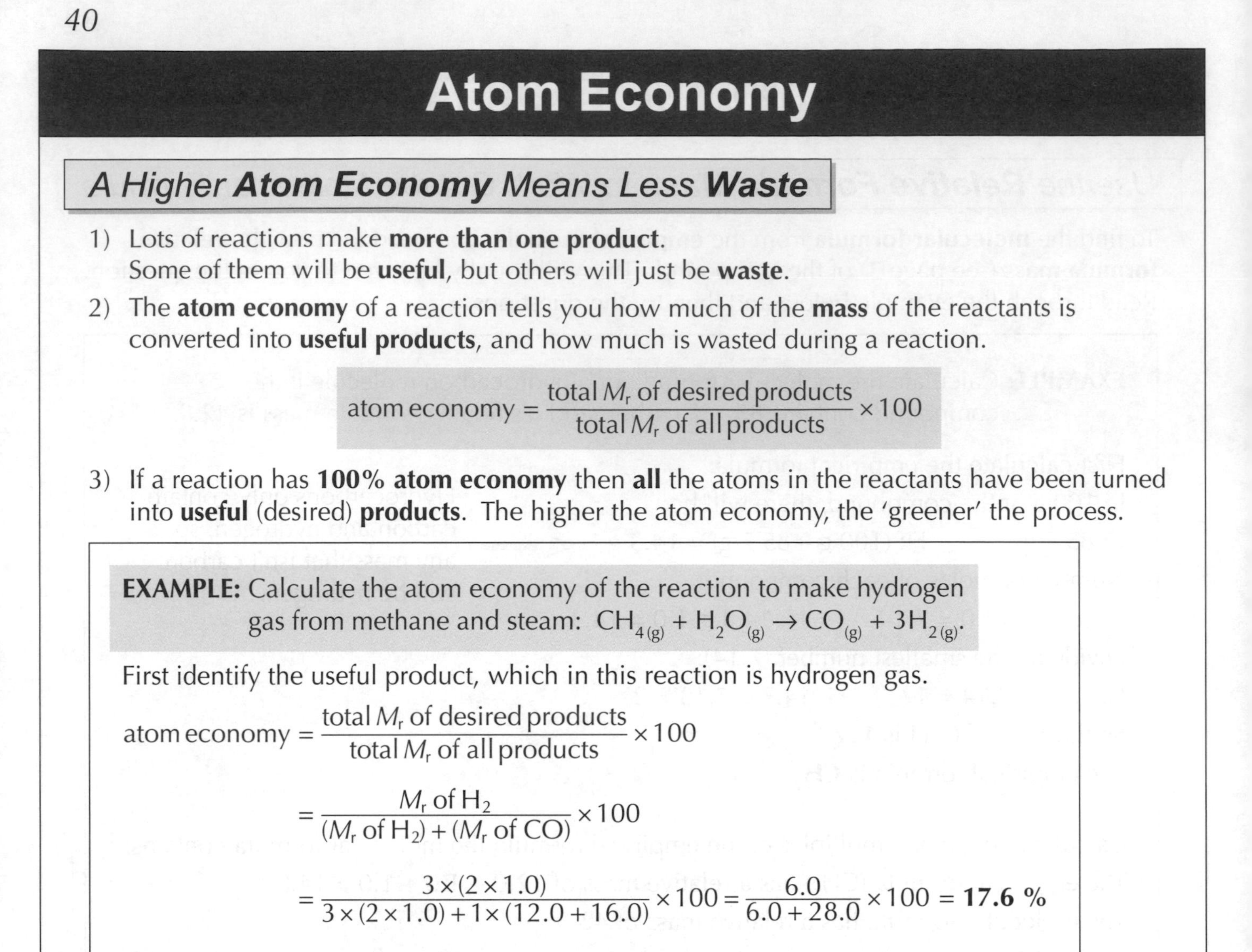

A Higher **Atom Economy** Means Less **Waste**

1) Lots of reactions make **more than one product**. Some of them will be **useful**, but others will just be **waste**.
2) The **atom economy** of a reaction tells you how much of the **mass** of the reactants is converted into **useful products**, and how much is wasted during a reaction.

$$\text{atom economy} = \frac{\text{total } M_r \text{ of desired products}}{\text{total } M_r \text{ of all products}} \times 100$$

3) If a reaction has **100% atom economy** then **all** the atoms in the reactants have been turned into **useful** (desired) **products**. The higher the atom economy, the 'greener' the process.

EXAMPLE: Calculate the atom economy of the reaction to make hydrogen gas from methane and steam: $CH_{4(g)} + H_2O_{(g)} \rightarrow CO_{(g)} + 3H_{2(g)}$.

First identify the useful product, which in this reaction is hydrogen gas.

$$\text{atom economy} = \frac{\text{total } M_r \text{ of desired products}}{\text{total } M_r \text{ of all products}} \times 100$$

$$= \frac{M_r \text{ of } H_2}{(M_r \text{ of } H_2) + (M_r \text{ of } CO)} \times 100$$

$$= \frac{3 \times (2 \times 1.0)}{3 \times (2 \times 1.0) + 1 \times (12.0 + 16.0)} \times 100 = \frac{6.0}{6.0 + 28.0} \times 100 = \mathbf{17.6\ \%}$$

High Atom Economy is Better in Industry

1) **Industrial reactions** are designed to be as **cheap** and **green** as possible. Generally, reactions with high atom economies are the **most efficient** processes as there is **minimal waste**.
2) The reactions with the **highest** atom economy are the ones that only have **one product**. These reactions have an atom economy of **100%**.
3) Reactions with low atom economies **use up resources** very quickly. They also make lots of **waste** materials that have to be **disposed** of somehow. That tends to make these reactions **unsustainable** — the raw materials run out and the waste has to go somewhere.
4) For the same reasons, low atom economy reactions aren't usually **profitable**. Raw materials are **expensive to buy**, and waste products can be expensive to **dispose of**.
5) The best way around the problem is to find a **use** for the waste products or to find a reaction with a **better** atom economy to make the same product.

Atom (Economy) — upgrade to Superior for only £16.99...

1) a) Ethanol can be made from bromoethane in the following reaction:
$CH_3CH_2Br + NaOH \rightarrow CH_3CH_2OH + NaBr$.
What is the atom economy of this reaction?

b) In industry, ethanol is made from ethene and steam using the following reaction:
$CH_2CH_2 + H_2O \rightarrow CH_3CH_2OH$.
Suggest why this reaction is used, rather than the reaction in part a).

Endothermic and Exothermic Reactions

In an **exothermic** reaction, **heat** energy is **given out** (the room temperature rises).
In an **endothermic** reaction, **heat** energy is **taken** from the surroundings (the room temperature drops).

***Making** and **Breaking** Bonds*

1) It takes energy to **break bonds**. When two atoms joined by a bond are **separated**, the energy required to do this must be provided from the surroundings.
2) However, energy is **released** when bonds are made. When two atoms become **joined together** by forming a bond, energy is **released** to the surroundings.
3) In a reaction, if more energy is taken in to break bonds than is given out when bonds are made, the process is **endothermic** — it will take in heat energy. The overall **enthalpy change** of the reaction (ΔH) is **positive**.
4) But, if more energy is given out when bonds are made than is taken in when bonds are broken, the process is **exothermic** — it will give out heat energy. The overall **enthalpy change** of an exothermic reaction (ΔH) is **negative**.

*Reactions can be Represented by **Energy Level Diagrams***

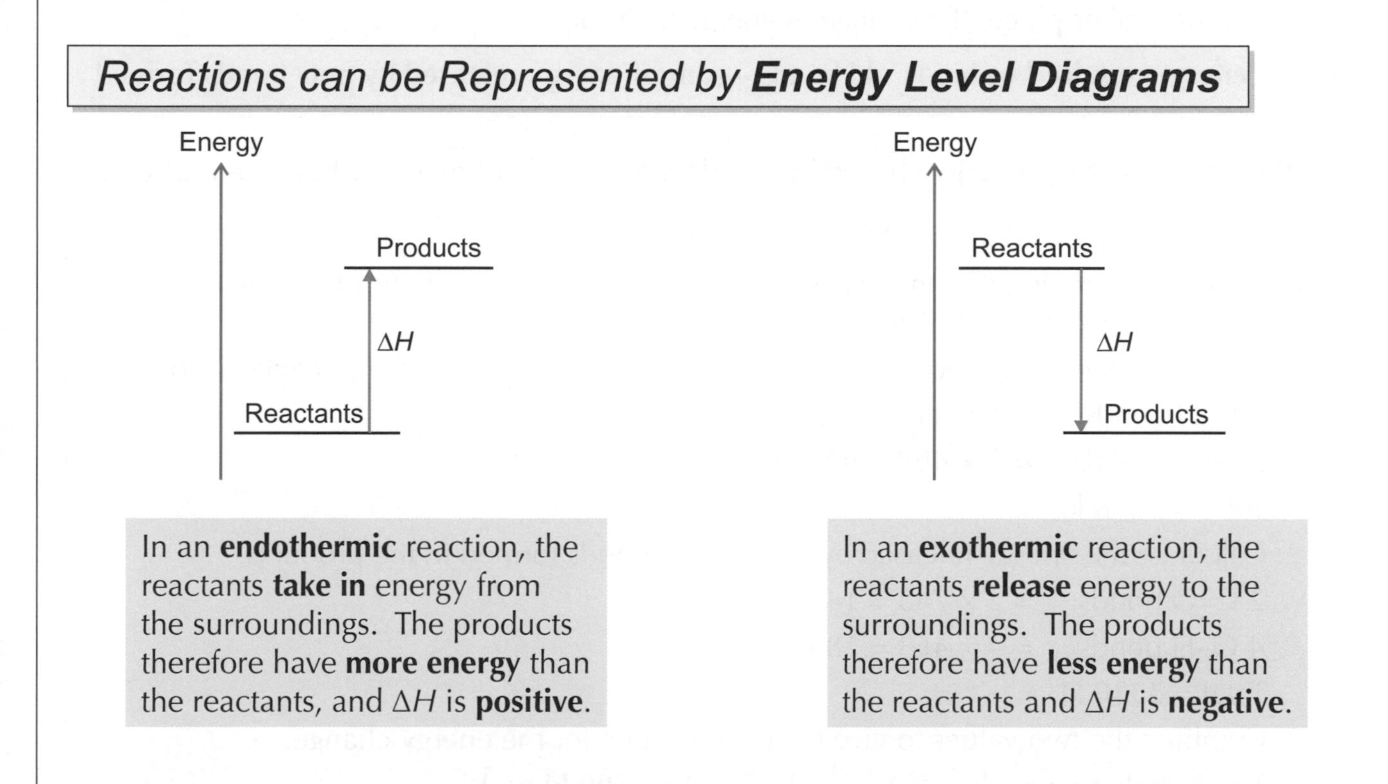

In an **endothermic** reaction, the reactants **take in** energy from the surroundings. The products therefore have **more energy** than the reactants, and ΔH is **positive**.

In an **exothermic** reaction, the reactants **release** energy to the surroundings. The products therefore have **less energy** than the reactants and ΔH is **negative**.

After that I think I need a cup of tea. It'll help improve my energy level...

1) Are the following reactions exothermic or endothermic?
   a) burning coal
   b) sodium hydrogencarbonate + hydrochloric acid (temperature drops)
   c) acid + hydroxide (gets hotter)
   d) methane + steam (cools as they react)
2) a) Draw an energy level diagram for the following reaction:
   $C_6H_{12}O_6 + 6O_2 \rightarrow 6CO_2 + 6H_2O$ $\quad \Delta H = -2809 \text{ kJ mol}^{-1}$
   You should label the products, reactants and enthalpy change on your diagram.
   b) Is the reaction in part a) endothermic or exothermic?

Bond Energy

Average Bond Energy

Bonds between **different atoms** require different amounts of **energy** to break them. When the **same two atoms** bond in the same way, the amount of energy needed is always about the same. The average bond energy values for some common bonds are given below:

C–H 413	C–O 360	C=C 612
O=O 498	H–H 436	C=O 743
C–C 348	O–H 463	

All these values are in kJ mol^{-1}.

The values tell you that:

e.g. It takes 413 kJ of energy to break 1 mole of C–H bonds.

It takes 463 × 2 = 926 kJ to break 1 mole of water (which has 2 O–H bonds per molecule) into oxygen and hydrogen atoms.

743 × 2 = 1486 kJ are released when 1 mole of CO_2 (which has 2 C=O bonds) forms.

Calculating the *Change* in Energy

When a reaction takes place, the change in energy is simply:

sum of energy required to break old bonds – sum of energy released by new bonds formed

EXAMPLE: Calculate the energy change involved when 1 mole of methane burns in oxygen.

The equation for the reaction is: $CH_4 + 2O_2 \rightarrow CO_2 + 2H_2O$

This tells you that 1 mole of methane reacts with 2 moles of oxygen to form 1 mole of carbon dioxide and 2 moles of water.

Step 1: Calculate the energy required to break all of the bonds between the reactant atoms:

4 C–H bonds = 4 × 413 = 1652 kJ

2 O=O bonds = 2 × 498 = 996 kJ

Total = 2648 kJ

Step 2: Calculate the energy released by all the new bonds formed in the products:

2 C=O bonds = 2 × 743 = 1486 kJ

4 O–H bonds = 4 × 463 = 1852 kJ

Total = 3338 kJ

Step 3: Combine the two values to give the overall value for the energy change:

The overall energy change is: 2648 – 3338 = **–690 kJ mol^{-1}**.

The negative sign shows that energy is being released to the surroundings, indicating that this is an **exothermic** reaction. This is expected, since this is a combustion reaction.

Ian Fleming was like an exothermic reaction — he made lots of Bonds...

1) Calculate the energy change of the following reactions:
(Use the values for the average bond energies given at the top of the page).

a) burning 1 mole of propane $C_3H_8 + 5O_2 \rightarrow 3CO_2 + 4H_2O$

b) burning 1 mole of ethanol $C_2H_5OH + 3O_2 \rightarrow 2CO_2 + 3H_2O$

c) hydrogenation of 1 mole of ethene $C_2H_4 + H_2 \rightarrow C_2H_6$

Planning Experiments

Make Sure You Plan Your Experiment Carefully

To get accurate and precise results from your experiments, you first need to plan them carefully...

1) Work out the **aim** of the experiment.
2) Identify the **variables** (see below).
3) Decide what **data** to collect.
4) Decide the right **equipment** to use.
5) Plan how to reduce any **risks** in your experiment.
6) Write out a **detailed method.**
7) Carry out **tests** to address the aim of your experiment.

You Need to Control All the Variables

A **variable** is a quantity that might **change** during an experiment, for example temperature. There are two types of variables to know about when carrying out an experiment:

- The **independent variable** is the quantity that you **change**.
- The **dependent variable** is the thing that you measure.

When you plan an experiment you need to work out how you will **control** the variables so that the only one that changes is the one you're investigating — all the others are kept **constant**.

EXAMPLE: Measuring the effect of surface area on reaction rate.

In this experiment, the **independent variable** is the **surface area**, and the **dependent variable** is the **rate** of reaction.
Everything else, such as temperature and concentration, has to stay exactly the same between different experiments. Surface area is the only variable that you change.

Choose the Right Equipment

You need to think carefully about selecting the right **equipment** for your experiment...

1) The equipment has to be **appropriate** for the specific experiment — for example, in an experiment where you're collecting a **gas** the equipment you use needs to be properly **sealed** so that the gas can't **escape**.
2) The equipment needs to be the right **size**.
3) The equipment needs to be the right level of **sensitivity** — for example, if you want to measure out 4.2 g of a compound, you'll need a balance that measures to at least the nearest 0.1 g, not the nearest gram.

Reduce Risk — and play poker instead...

1) A student is measuring the effect of temperature on the time taken for a lump of magnesium to react completely in a sample of concentrated hydrochloric acid.
   a) What is the dependent variable in the student's experiment?
   b) Name two variables that the student should control to make the experiment a fair test.

Presenting and Interpreting Data

You Can Represent Your Data in a **Table** or on a **Graph**

When you do an experiment, it's a good idea to set up a table to **record** your **results** in. Make sure you **include** enough **rows** and **columns** to **record all of the data** you need. Tables are good for **recording** data, but it can be easier to interpret your results if you **plot** them on a **graph**. Depending on the **type** of experiment, the **graph** you plot will vary:

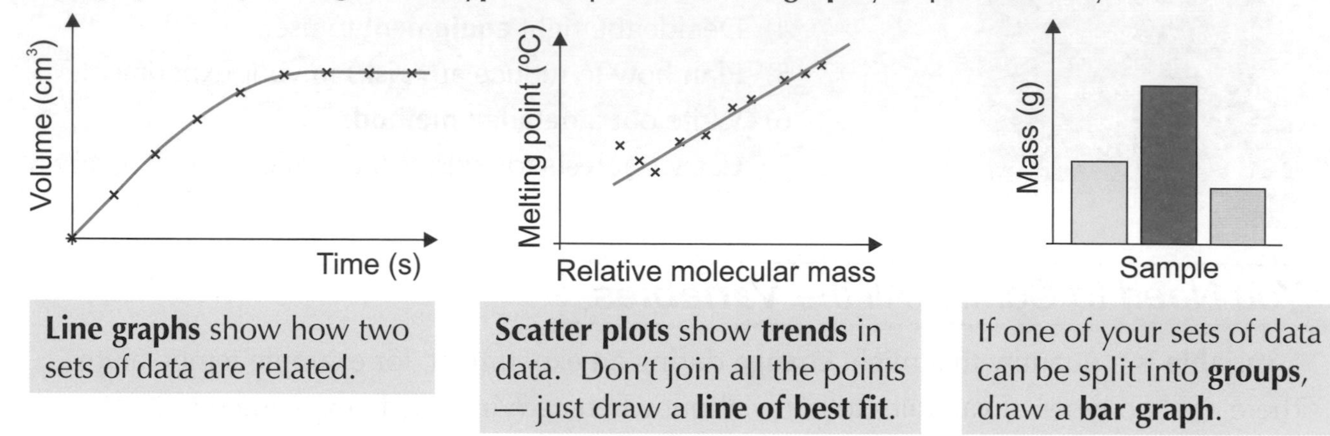

Line graphs show how two sets of data are related.

Scatter plots show **trends** in data. Don't join all the points — just draw a **line of best fit**.

If one of your sets of data can be split into **groups**, draw a **bar graph**.

Repeating an Experiment Makes Your Results More **Reliable**

1) If you **repeat** an experiment, your results will usually **differ slightly** each time you do it. You can use the **mean** (or average) of the measurements to represent all these values. The more times you repeat the experiment the **more reliable** the average will be. To find the mean:

Add together all the data values then **divide** by the total number of values in the sample.

EXAMPLE: Calculate the mean result for the volume of hydrogen gas produced after 30 seconds in the reaction between hydrochloric acid and magnesium.

Run 1	Run 2	Run 3
23 cm^3	22 cm^3	25 cm^3

There are **three** values in this sample, so to find the mean result, just add together the results and divide by three:

$(23 + 22 + 25) \div 3 = \mathbf{23.3\ cm^3}$

2) Repeating experiments also lets you spot any **weird results** that stick out like a hedgehog in a tea cup. These are called **anomalous** results. For example — if one of the results above was only 5 cm^3, then something probably went wrong. You should **ignore** the anomalous result when you calculate the mean.
3) Anomalous results are really easy to spot on **scatter plots** and **line graphs** as they sit miles away from the line of best fit.

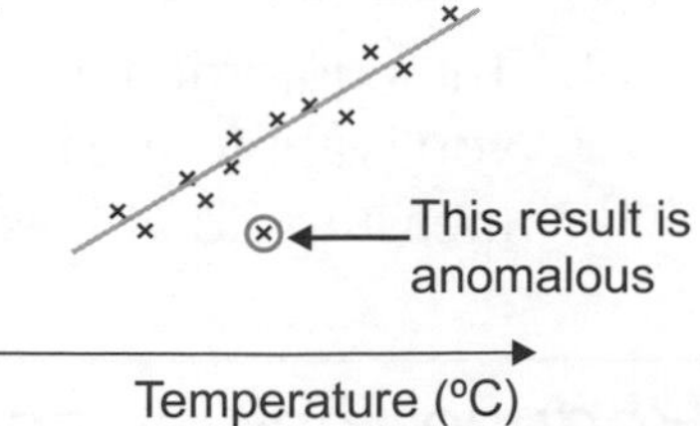

I was hoping for a nice result, but it ended up being mean...

1) Kay measured the volume of gas given off in a reaction. Her results were 22.0 cm^3, 23.0 cm^3, 22.0 cm^3, 19.0 cm^3 and 24.0 cm^3. Identify any anomalous results and calculate the mean.

Conclusions and Error

Measurements Always Have **Uncertainty** in Them

1) You saw on the last page that if you **repeat** an experiment then the results normally **vary** slightly, even if you do everything **exactly the same** each time. This is partly to do with the fact that there will always be a degree of **error** in any measurements you make.
   For example, if you use **measuring scales** that measure to the nearest **gram**, then whenever you weigh something, you might have **up to 0.5 g** more or less than the measured mass.
2) The **uncertainty** of a measurement is just the **maximum error** there could be — so in the example above it's **0.5 g**.
3) The **percentage error** of a measurement is just another way of showing the uncertainty. You calculate it using the following formula:

$$\text{percentage error} = \frac{\text{uncertainty}}{\text{measurement}} \times 100$$

EXAMPLE: Nikhil measures out 6.0 cm^3 of water in a measuring cylinder that has markings every 0.5 cm^3. Calculate the percentage error of his measurement.

The measuring cylinder can be read to the nearest 0.5 cm^3, so any measurement could be up to 0.25 cm^3 more than or less than the reading. So the uncertainty is 0.25 cm^3.

$$\text{Percentage error} = \frac{\text{uncertainty}}{\text{measurement}} \times 100 = \frac{0.25\ \text{cm}^3}{6.0\ \text{cm}^3} \times 100 = \mathbf{4.2\%}$$

Don't **Jump to Conclusions**

1) Collecting reliable data is important, but if the data doesn't answer the aim, it's not much use.
2) The **data** should **support** the conclusion. This may sound obvious, but it's easy to **jump** to conclusions. Conclusions should be **specific** — not make sweeping generalisations.
3) Conclusions often try to link changes in one variable with another. For your conclusion to be valid, you have to make sure all the other variables in the experiment were **controlled**.
4) And remember — **correlation doesn't** always mean **cause**.

EXAMPLE: Investigating whether chlorinated drinking water increases cancer risk.

Some studies claim that drinking chlorinated tap water increases the risk of some cancers. But it's hard to **control** all the **variables** between people who drink tap water and people who don't, so designing a fair test is very tricky.

But, even if some studies show a group of people who drink more chlorinated water are slightly more likely to get certain cancers, it doesn't mean that drinking chlorinated water **causes** cancer. There will be heaps of other differences between the groups of people. It could be due to any of them.

You've reached the conclusion. Please travel safely and mind the gap...

1) Calculate the percentage error in each of the following measurements:
   a) A mass of 1.4 g on a weighing scale that can measure to the nearest 0.1 g.
   b) A time of 23 seconds on a clock that times to the nearest second.
   c) A temperature of 10.6 °C on a thermometer that has markings every 0.2 °C.

Answers

Section 1 — The Structure of the Atom

Page 1 — Atomic Structure

1 Protons and neutrons.

2 +2

3 −2

Page 2 — Atomic Number, Mass Number and Isotopes

1 $A - Z = 31 - 15 = \mathbf{16}$

2 Two isotopes of the same element have the same number of protons and electrons but different numbers of neutrons.

3 All three isotopes have 6 protons and 6 electrons. Carbon-12 has (12 – 6) = 6 neutrons, carbon-13 has (13 – 6) = 7 neutrons and carbon-14 has (14 – 6) = 8 neutrons.

Page 3 — Relative Atomic Mass

1 $[(8 \times 6) + (92 \times 7)] \div 100 = \mathbf{6.92}$

2 $[(99 \times 12) + (1 \times 13)] \div 100 = \mathbf{12.01}$

3 $[(52 \times 107) + (48 \times 109)] \div 100 = \mathbf{107.96}$

4 $23.0 + 19.0 = \mathbf{42.0}$

5 $12.0 + (3 \times 1.0) + 35.5 = \mathbf{50.5}$

Page 4 — Electronic Structure

1

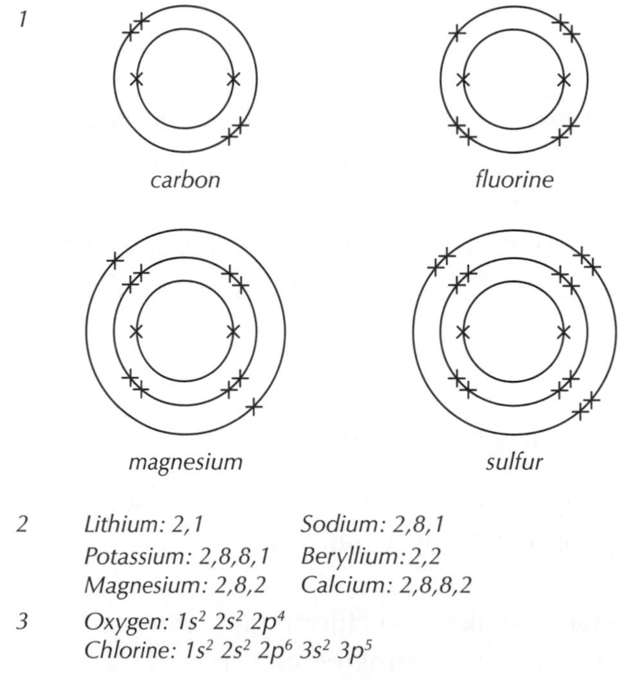

2 Lithium: 2,1 Sodium: 2,8,1
Potassium: 2,8,8,1 Beryllium: 2,2
Magnesium: 2,8,2 Calcium: 2,8,8,2

3 Oxygen: $1s^2\ 2s^2\ 2p^4$
Chlorine: $1s^2\ 2s^2\ 2p^6\ 3s^2\ 3p^5$

Page 5 — The Periodic Table

1

s-block elements	p-block elements
caesium	phosphorus
potassium	aluminium
calcium	sulfur
barium	

2 E.g. They have the same number of electrons in their outer shell. / They react in similar ways.

Section 2 — Formation of Ions

Page 6 — Ionisation Energy

1 $Na_{(g)} \rightarrow Na^{+}_{(g)} + e^{-}$

2 Nuclear charge, the distance of the electron from the nucleus and shielding by inner electrons.

3 Magnesium
Fluorine
Oxygen

Page 7 — Formation of Ions

1 +1

2 Group 7

3 SO_3^{2-}

Page 8 — Oxidation Numbers

1 The oxidation number tells you how many electrons an atom has donated or accepted.

2 Al^{3+}: +3 H^{+}: +1 Ne: 0 O^{2-}: −2

3 0

Section 3 — Intermolecular Bonding

Page 9 — Intermolecular Bonding

1 Weak intermolecular forces
Cl—Cl Cl—Cl Cl—Cl
Strong covalent bond

2 The trend should show an increase in boiling points as size of the alkane increases.
E.g. pentane: 36 °C, hexane: 69 °C, heptane: 98 °C, octane: 126 °C.

Page 10 — Polarity

1 a) HF as it is polar, and H_2 is non-polar.
b) H_2O as it can form hydrogen bonds, and H_2S can't.
c) CH_3F as fluorine is much more electronegative than carbon so it will be a polar molecule. Iodine is less electronegative / iodine and carbon have similar electronegativities, so CH_3I is non-polar.

2 E.g.

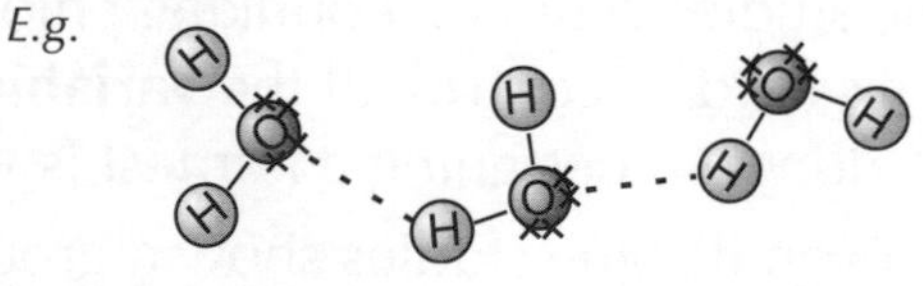

Section 4 — Bonding and Properties

Page 11 — Ionic Bonding

1

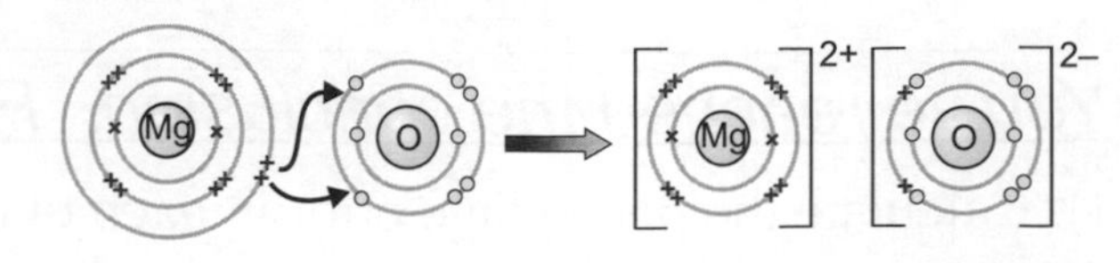

2 You need two K^{+} ions (2 × +1) to balance out each O^{2-} ion (1 × –2), so the ratio is 2:1. The ionic formula is K_2O.

Answers

Page 12 — Ionic Compounds

1 BeO, Li_2O, LiF
The higher the charges on the ions, the stronger the bonds between them. The stronger the bonds, the higher the melting point. Beryllium oxide is formed from ions which both have charges of magnitude 2. Lithium oxide is formed from oxide ions with a –2 charge and lithium ions with only a +1 charge. The ions in lithium fluoride both have charges of magnitude one. Therefore the strongest bonds will be in beryllium oxide, followed by lithium oxide, then lithium fluoride.

2 When potassium chloride is solid, the K^+ and Cl^- ions are held together in an ionic lattice, so they're not free to move and conduct electricity. When molten or dissolved, the ions separate, so they're free to move and able to carry a current.

Page 13 — Covalent Bonding

1 a) Chlorine: b) Water: c) Ethane: d) Oxygen:

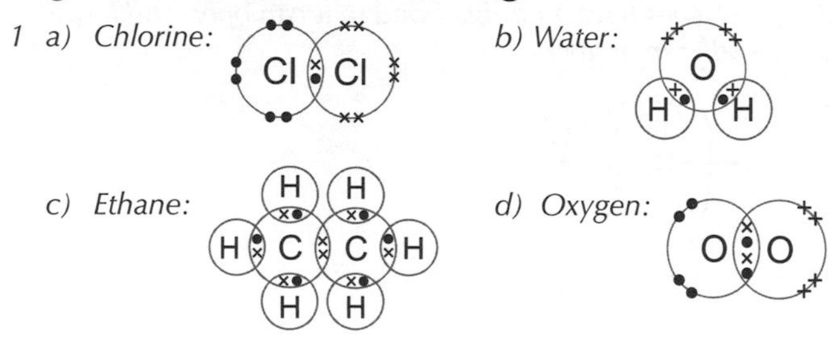

Page 14 — Small Covalent Molecules

1

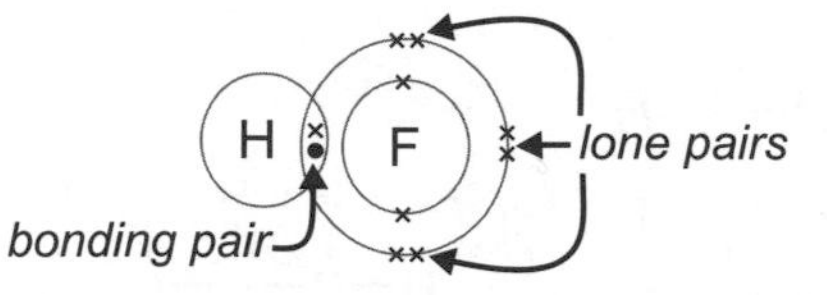

2 For a small covalent compound to boil, you have to break the weak intermolecular bonds between the molecules rather than the strong covalent bonds between the atoms in a molecule. These intermolecular bonds don't need much energy to break, so nitrogen has a low boiling point and is a gas at room temperature.

Page 15 — Giant Covalent Structures

1 You could use solubility, though not all ionic compounds are soluble in water so this test may prove inconclusive. Melt and check to see if the molten substance conducts electricity— if it does it's probably ionic. You need to be careful of graphite though, which is a giant covalent molecule that is able to conduct electricity. To get around this, you could test the conductivity of the crystal in its solid form as well. While solid graphite will conduct, ionic salts only conduct electricity when molten or in solution.

2 In sodium chloride, the energy required to break the strong ionic bonds is provided when the ions become surrounded by water molecules. Fewer strong bonds are replaced by many more weaker bonds. In the case of a giant covalent molecule there is no way to get the energy required to break the many strong covalent bonds between atoms, so diamond doesn't dissolve.

Page 16 — Metallic Bonding

1 Calcium is likely to have a higher melting point. This is because calcium is made up of a lattice of Ca^{2+} ions, with two delocalised electrons per ion. Potassium however is made up of K^+ ions and only one delocalised electron per ion. So the bonding in calcium is stronger, and will require more energy to be broken leading to a higher melting point.

2

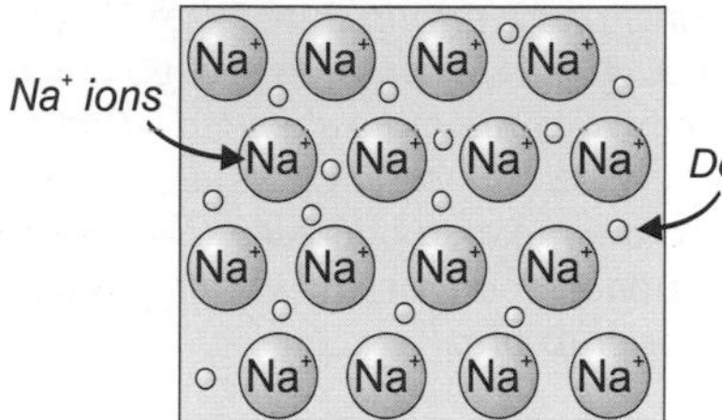

3 Similarities, e.g. high melting and boiling points / both conduct electricity when molten.
Differences, e.g. solid sodium chloride is soluble in water, sodium is not / solid sodium chloride is an electrical insulator but solid sodium conducts electricity.

Page 17 — Trends in Properties Across the Periodic Table

1 The melting points of the oxides of sodium, magnesium and aluminium are all high. Also, all three are conductors of electricity when molten. These properties clearly point towards the oxides being ionic.
Silicon dioxide also has a high melting point but is a non-conductor when molten so it has a giant covalent structure.
Oxides of phosphorus and sulfur have low melting points and are non-conductors. These are therefore likely to be small covalent molecules.

2 The melting points will be high up to silicon (which forms strong covalent bonds) then drop down to phosphorus and sulfur which are small covalent molecules.

3 a) Sodium chloride has ionic bonds.
b) The chloride of phosphorus has covalent bonds and is a small covalent molecules.

Section 5 — Chemical Equations

Page 19 — Writing and Balancing Equations

1 Step 2: $CH_4 + O_2 \rightarrow CO_2 + H_2O$
Step 3: $CH_4 + 2O_2 \rightarrow CO_2 + 2H_2O$
(The Cs already balance, so put a 2 in front of H_2O to balance the Hs. Now put a 2 in front of O_2 to balance the Os. Check that all still balances.)

2 a) $C_2H_5OH + 3O_2 \rightarrow 2CO_2 + 3H_2O$
b) $Ca(OH)_2 + 2HCl \rightarrow CaCl_2 + 2H_2O$

3 First balance the atoms:
$Cl_{2(g)} + Fe^{2+}_{(aq)} \rightarrow 2Cl^-_{(aq)} + Fe^{3+}_{(aq)}$
Then balance the charges:
$Cl_{2(g)} + 2Fe^{2+}_{(aq)} \rightarrow 2Cl^-_{(aq)} + 2Fe^{3+}_{(aq)}$

Answers

Section 6 — Inorganic Chemistry

Page 20 — Group 2

1 *Reactivity increases down Group 2, so you'd expect magnesium to react very slowly with cold water to produce hydrogen gas and magnesium hydroxide. You'd expect strontium to react vigorously with cold water to produce hydrogen gas and strontium hydroxide.*

2 *The boiling points of the Group 2 metals will decrease down the group. This is because, as you go down the Group, the nuclei become more shielded. This causes the attraction between the positive metal ions and the free electrons in the metal to decrease. So the strength of the metallic bonds decreases down the group, making them easier to break.*

Page 21 — Group 7

1 a) *Chlorine is more reactive than bromine, so it displaces the bromide in solution producing chloride and bromine.*

b) *Iodine is less reactive than chlorine, so no reaction takes place.*

c) *Iodine is less reactive than bromine, so no reaction takes place.*

d) *Chlorine is more reactive than iodine, so it displaces the iodide in solution producing chloride and iodine.*

2

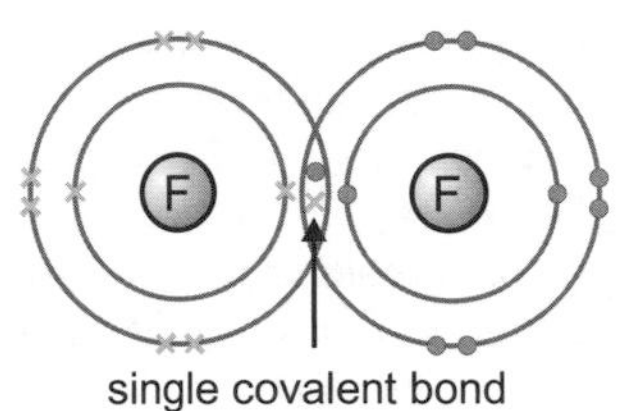

Page 22 — Acids and Bases

1 $HNO_3 + KOH \rightarrow KNO_3 + H_2O$

2 a) $H_2SO_4 \xrightarrow{water} 2H^+ + SO_4^{2-}$

b) $KOH \xrightarrow{water} K^+ + OH^-$

c) $HNO_3 \xrightarrow{water} H^+ + NO_3^-$

Section 7 — Organic Chemistry

Page 23 — Organic Molecules

1 *skeletal:* *displayed:*

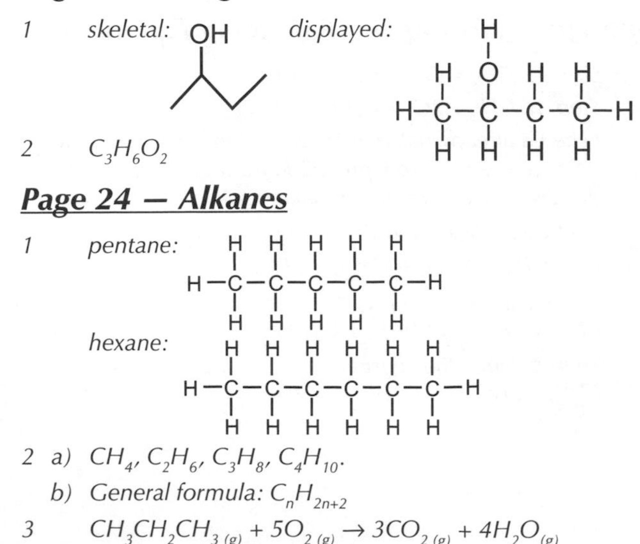

2 $C_3H_6O_2$

Page 24 — Alkanes

1 *pentane:*

hexane:

2 a) CH_4, C_2H_6, C_3H_8, C_4H_{10}.

b) *General formula:* C_nH_{2n+2}

3 $CH_3CH_2CH_{3\,(g)} + 5O_{2\,(g)} \rightarrow 3CO_{2\,(g)} + 4H_2O_{(g)}$

Page 25 — Alkenes

1 *e.g.*

2 *Two from:*

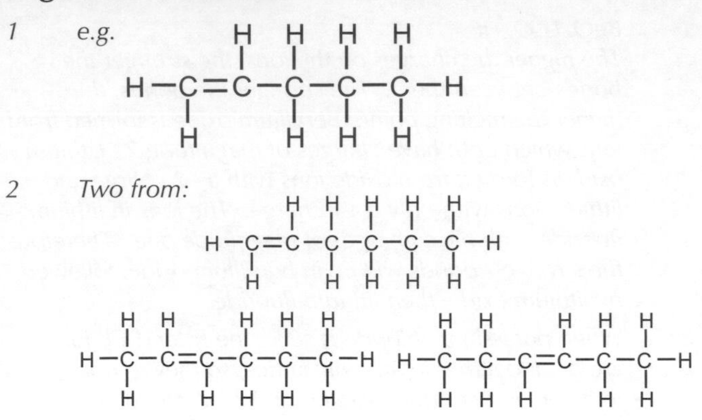

3 *General formula:* C_nH_{2n}

Page 26 — Polymerisation

1 *Alkenes have a double bond that can open and link to other monomers.*

2 a)

b)

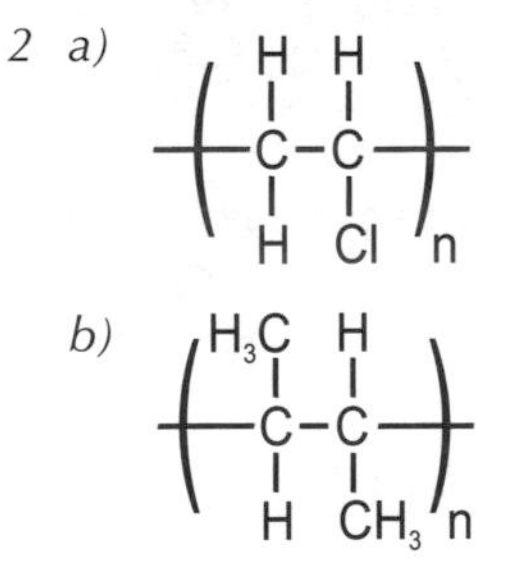

Page 27 — Alcohols

1

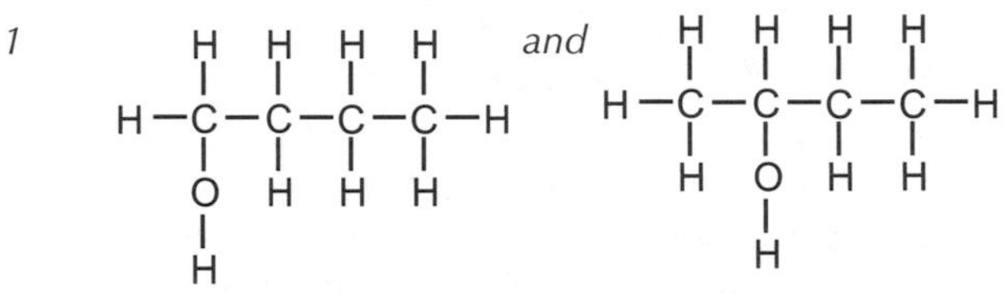

2 *General formula:* $C_nH_{(2n+1)}OH$

3 *Ethanol will have a higher melting point than ethane because it is able to form hydrogen bonds. These require more energy to break than the intermolecular bonds that form between nonpolar molecules such as ethane.*

Section 8 — Chemical Reactions

Page 30 — Reaction Types

1 a) *combustion, exothermic, oxidation, redox, (reduction)*

b) *displacement, oxidation, precipitation, redox, reduction, substitution*

c) *exothermic, neutralisation, oxidation, substitution, redox, reduction*

d) *addition, hydrogenation, (oxidation, redox, reduction)*

Answers

Section 9 — Rates of Reaction

Page 31 — Reaction Rates

1 a) e.g. Measure the change in temperature / Measure the change in pH / Measure the loss of mass as CO_2 is evolved / Measure the volume of CO_2 produced using a gas syringe.

b) e.g. Measure how long it takes for the solution to go cloudy / Measure the loss of mass as SO_2 is evolved / Measure the volume of SO_2 produced using a gas syringe.

c) e.g. Measure the change in pH / Measure the loss of mass as CO_2 is evolved / Measure the volume of CO_2 produced using a gas syringe.

Page 32 — Collision Theory

1 The particles may have collided facing in the wrong direction or without enough energy to react.

2 The activation energy is the minimum amount of kinetic energy particles need to react.

3

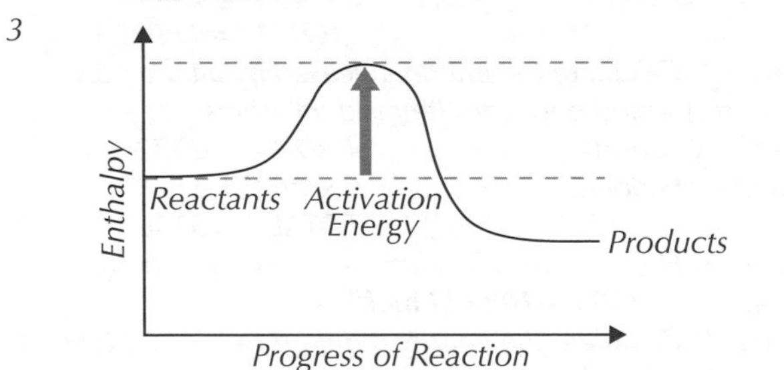

Page 33 — Reaction Rates and Catalysts

1 e.g. Increase the temperature / Increase the concentration of the reactants / Add a catalyst.

2 The particles in the gas will be closer together so they're more likely to collide.

3 A catalyst is a substance that speeds up the rate of reaction without being changed or used up itself.

4 E.g. Catalysts make reactions cheaper to run and reduce their CO_2 emissions.

Section 10 — Equilibria

Page 34 — Reversible Reactions

1 a) At the start of the reaction, the rate of the forward reaction is faster than the rate of the backwards reaction.

b) At equilibrium, the rates of the forward and backward reactions are the same.

2 Dynamic equilibrium is where the forwards and backwards reactions of a reversible reaction are going at the same rate. This means that although both reactions are still happening, the concentrations of the products and reactants don't change.

Page 35 — Le Chatelier's Principle

1 Increasing the amount of steam will increase the concentration of particles on the left of the equation (which will also increase the pressure on the left hand side), and move the position of equilibrium to the right, increasing the yield of ethanol.

2 a) The equilibrium will move to the left to favour the endothermic reaction.

b) The equilibrium will move to the right to try and increase the concentration of ammonia.

Page 36 — Equilibrium and Yield

1 The temperature is low, which would favour the forward reaction, and increase the yield of ethanol.
But it is so low that the forward reaction rate will be much too slow to be economic.

2 The pressure is high, which would favour the forward reaction, and increase the yield of ethanol.
But such a high pressure would be very expensive to maintain, making the reaction uneconomic.

Section 11 — Calculations

Page 37 — The Mole

1 molar mass = mass ÷ moles
= 68.6 ÷ 0.700 = **98.0 g mol^{-1}**

2 Molar mass NaCl = 23.0 + 35.5 = 58.5 g mol^{-1}
moles = mass ÷ molar mass = 117 ÷ 58.5 = **2.00 moles**

3 Molar mass water = 16.0 + (2 × 1.0) = 18.0 g mol^{-1}
moles of water = 54.0 ÷ 18.0 = 3.00 moles.
Molar mass of iron = 55.8 g mol^{-1}
moles of iron = 84.0 ÷ 55.8 = 1.51 moles.
There are more moles of water.

Page 38 — Determination of Formulae from Experiments

1 a) Mass of each substance: Pb: 12.9 g O: 1.00 g
Number of moles of each substance:
Pb: 12.9 ÷ 207.2 = 0.0623 moles
O: 1.00 ÷ 16.0 = 0.0625 moles
Divide by the smallest number (0.0623):
Pb: 1.00 O: 1.00
The ratio of Pb : O is 1 : 1. The empirical formula is **PbO**.

b) Mass of each substance: Fe: 2.33 g O: 1.00 g
Number of moles of each substance:
Fe: 2.33 ÷ 55.8 = 0.0418 moles
O: 1.00 ÷ 16 = 0.0625 moles
Divide by the smallest number (0.0418):
Fe: 1.00 O: 1.50
Multiply by two to give whole numbers: Fe: 2 O: 3
The ratio of Fe : O is 2 : 3. The empirical formula is $\mathbf{Fe_2O_3}$.

2 Percentage composition of each substance:
H: 4.30% C: 26.1% O: 69.6%
Number of moles of each substance:
H: 4.3 ÷ 1.0 = 4.30 moles
C: 26.1 ÷ 12.0 = 2.18 moles
O: 69.6 ÷ 16.0 = 4.35 moles
Divide by the smallest number (2.18):
H: 1.97 C: 1.00 O: 2.00
The ratio of H : C : O is 2 : 1 : 2.
The empirical formula is $\mathbf{CH_2O_2}$.

Answers

Page 39 — Calculation of Molecular Formulae

1 Mass of each substance in 100 g:
C: 52.2 g H: 13.0 g O: 34.8 g
Number of moles of each substance:
C: 52.2 ÷ 12.0 = 4.35 moles
H: 13.0 ÷ 1.0 = 13.0 moles
O: 34.8 ÷ 16.0 = 2.18 moles
Divide by the smallest number (2.18):
C: 2.00 H: 5.96 O: 1.00
Ratio of C : H : O is 2 : 6 : 1.
The empirical formula is C_2H_6O.
The empirical formula has a relative mass of (2 × 12.0) + (6 × 1.0) + (1 × 16.0) = 46.0. This is the same as the relative formula mass of the compound, so the molecular formula is also **C_2H_6O**.

2 Mass of each substance in 100 g:
C: 92.3 g H: 7.70 g
Number of moles of each substance:
C: 92.3 ÷ 12.0 = 7.69 moles
H: 7.70 ÷ 1.0 = 7.70 moles
Divide by the smallest number (7.69):
C: 1.00 H: 1.00
So the ratio of C : H is 1 : 1. The empirical formula is CH.
The empirical formula has a relative formula mass of 12.0 + 1.0 = 13.0.
78.0 ÷ 13.0 = 6.00, so there are six lots of the empirical formula in the compound.
The molecular formula is **C_6H_6**.

3 a) $\frac{1 \times 16.0}{(2 \times 12.0) + (6 \times 1.0) + (1 \times 16.0)} \times 100 = \mathbf{34.8\%}$

b) $\frac{3 \times 16.0}{(1 \times 1.0) + (1 \times 14.0) + (3 \times 16.0)} \times 100 = \mathbf{76.2\%}$

c) $\frac{1 \times 16.0}{(3 \times 12.0) + (6 \times 1.0) + (1 \times 16.0)} \times 100 = \mathbf{27.5\%}$

Page 40 — Atom Economy

1 a) M_r C_2H_5OH = (2 × 12.0) + (6 × 1.0) + 16.0 = 46.0
M_r NaBr = 23.0 + 79.9 = 102.9
$\frac{46.0}{46.0 + 102.9} \times 100 = \mathbf{30.9\%}$

b) This reaction has an atom economy of 100%, so will be cheaper and greener to run.

Section 12 — Enthalpy

Page 41 — Endothermic and Exothermic Reactions

1 a) exothermic
b) endothermic
c) exothermic
d) endothermic

2 a) Energy

$C_6H_{12}O_6 + 6O_2$

$\Delta H = -2809$ kJ mol^{-1}

$6CO_2 + 6H_2O$

b) exothermic

Page 42 — Bond Energy

1 a) Step 1: Calculate the energy required to break all of the bonds between the reactant atoms:
8 C — H bonds = 8 × 413 = 3304
2 C — C bonds = 2 × 348 = 696
5 O = O bonds = 5 × 498 = 2490
TOTAL = 6490
Step 2: Calculate the energy released by all the new bonds formed between the product atoms:
6 C = O bonds = 6 × 743 = 4458
8 O — H bonds = 8 × 463 = 3704
TOTAL = 8162
Step 3: Find the overall value for the energy change:
+6490 – 8162 = **–1672 kJ mol^{-1}**

b) Step 1: Calculate the energy required to break all of the bonds between the reactant atoms:
1 C — C bond = 1 × 348 = 348
1 C — O bond = 1 × 360 = 360
5 C — H bonds = 5 × 413 = 2065
1 O — H bonds = 1 × 463 = 463
3 O = O bond = 3 × 498 = 1494
TOTAL = 4730
Step 2: Calculate the energy released by all the new bonds formed between the product atoms
4 C = O bonds = 4 × 743 = 2972
6 O — H bonds = 6 × 463 = 2778
TOTAL = 5750
Step 3: Find the overall value for the energy change:
+4730 – 5750 = **–1020 kJ mol^{-1}**

c) Step 1: Calculate the energy required to break the H–H and C=C bonds:
1 H — H bond = 1 × 436 = 436
1 C = C bond = 1 × 612 = 612
TOTAL = 1048
Step 2: Calculate the energy released by all the new bonds formed between product atoms
2 C — H bonds = 2 × 413 = 826
1 C — C bond = 1 × 348 = 348
TOTAL = 1174
Step 3: Find the overall value for the energy change:
+1048 – 1174 = **–126 kJ mol^{-1}**

Section 13 — Investigating & Interpreting

Page 43 — Planning Experiments

1 a) The time taken for the lump of magnesium to react.
b) E.g. The mass of magnesium used / The concentration of hydrochloric acid / The volume of hydrochloric acid / The surface area of the magnesium.

Page 44 — Presenting and Interpreting Data

1 The anomalous result is 19.0 cm^3.
The mean is: (22.0 + 23.0 + 22.0 + 24.0) ÷ 4 = **22.8 cm^3**

Page 45 — Conclusions and Error

1 a) The uncertainty of the weighing scales is 0.05 g, so the percentage error = $\frac{0.05}{1.4} \times 100 = \mathbf{3.6\%}$

b) The uncertainty of the clock is 0.5 s, so the percentage error = $\frac{0.5}{23} \times 100 = \mathbf{2.2\%}$

c) The uncertainty of the thermometer is 0.1 °C, so the percentage error = $\frac{0.1}{10.6} \times 100 = \mathbf{0.9\%}$

Index

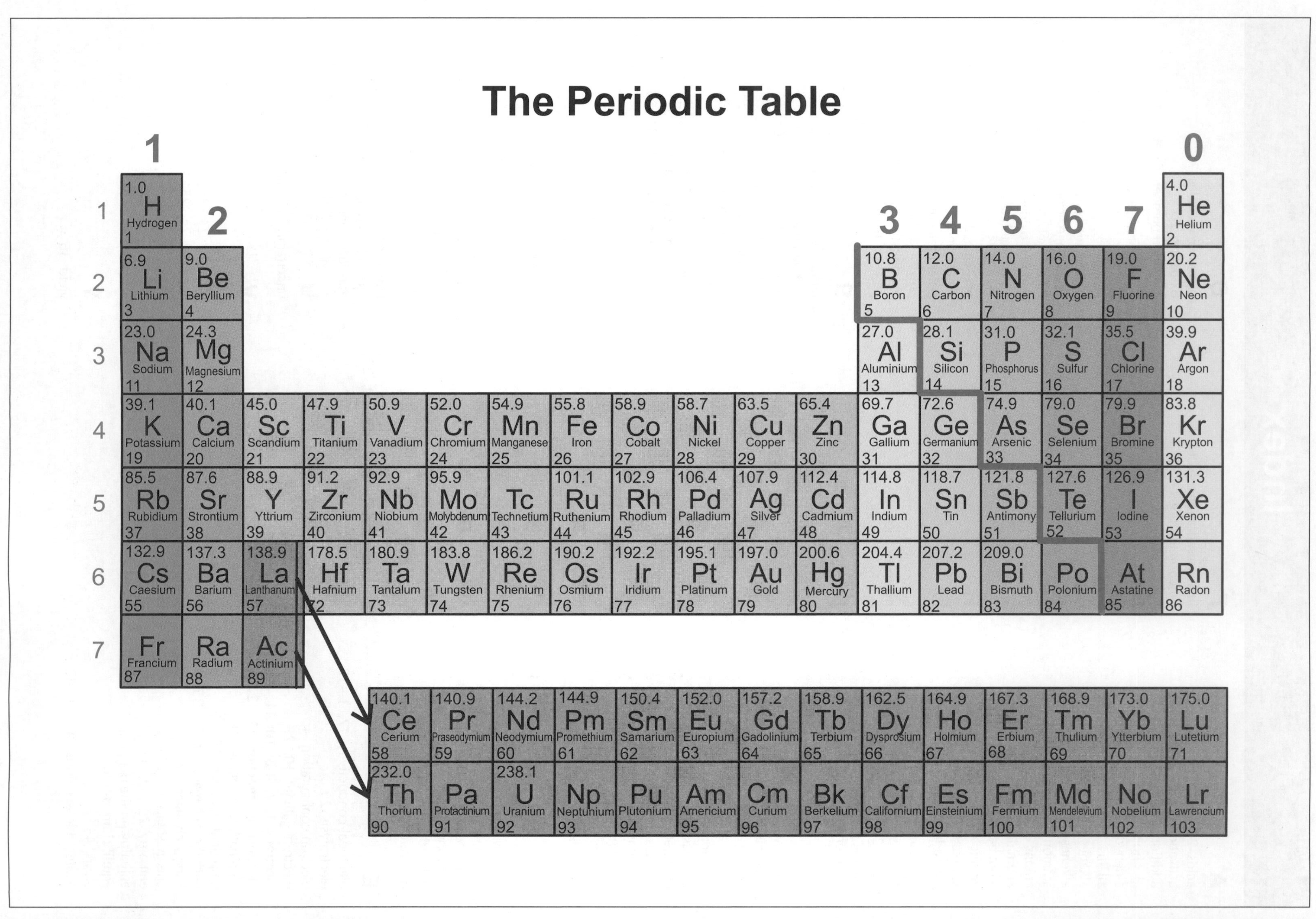
The Periodic Table
1
2
3
4
5
6
7
0
1
2
3
4
5
6
7
1.0 H Hydrogen 1
4.0 He Helium 2
6.9 Li Lithium 3
9.0 Be Beryllium 4
10.8 B Boron 5
12.0 C Carbon 6
14.0 N Nitrogen 7
16.0 O Oxygen 8
19.0 F Fluorine 9
20.2 Ne Neon 10
23.0 Na Sodium 11
24.3 Mg Magnesium 12
27.0 Al Aluminium 13
28.1 Si Silicon 14
31.0 P Phosphorus 15
32.1 S Sulfur 16
35.5 Cl Chlorine 17
39.9 Ar Argon 18
39.1 K Potassium 19
40.1 Ca Calcium 20
45.0 Sc Scandium 21
47.9 Ti Titanium 22
50.9 V Vanadium 23
52.0 Cr Chromium 24
54.9 Mn Manganese 25
55.8 Fe Iron 26
58.9 Co Cobalt 27
58.7 Ni Nickel 28
63.5 Cu Copper 29
65.4 Zn Zinc 30
69.7 Ga Gallium 31
72.6 Ge Germanium 32
74.9 As Arsenic 33
79.0 Se Selenium 34
79.9 Br Bromine 35
83.8 Kr Krypton 36
85.5 Rb Rubidium 37
87.6 Sr Strontium 38
88.9 Y Yttrium 39
91.2 Zr Zirconium 40
92.9 Nb Niobium 41
95.9 Mo Molybdenum 42
Tc Technetium 43
101.1 Ru Ruthenium 44
102.9 Rh Rhodium 45
106.4 Pd Palladium 46
107.9 Ag Silver 47
112.4 Cd Cadmium 48
114.8 In Indium 49
118.7 Sn Tin 50
121.8 Sb Antimony 51
127.6 Te Tellurium 52
126.9 I Iodine 53
131.3 Xe Xenon 54
132.9 Cs Caesium 55
137.3 Ba Barium 56
138.9 La Lanthanum 57
178.5 Hf Hafnium 72
180.9 Ta Tantalum 73
183.8 W Tungsten 74
186.2 Re Rhenium 75
190.2 Os Osmium 76
192.2 Ir Iridium 77
195.1 Pt Platinum 78
197.0 Au Gold 79
200.6 Hg Mercury 80
204.4 Tl Thallium 81
207.2 Pb Lead 82
209.0 Bi Bismuth 83
Po Polonium 84
At Astatine 85
Rn Radon 86
Fr Francium 87
Ra Radium 88
Ac Actinium 89
140.1 Ce Cerium 58
140.9 Pr Praseodymium 59
144.2 Nd Neodymium 60
144.9 Pm Promethium 61
150.4 Sm Samarium 62
152.0 Eu Europium 63
157.2 Gd Gadolinium 64
158.9 Tb Terbium 65
162.5 Dy Dysprosium 66
164.9 Ho Holmium 67
167.3 Er Erbium 68
168.9 Tm Thulium 69
173.0 Yb Ytterbium 70
175.0 Lu Lutetium 71
232.0 Th Thorium 90
Pa Protactinium 91
238.1 U Uranium 92
Np Neptunium 93
Pu Plutonium 94
Am Americium 95
Cm Curium 96
Bk Berkelium 97
Cf Californium 98
Es Einsteinium 99
Fm Fermium 100
Md Mendelevium 101
No Nobelium 102
Lr Lawrencium 103